ServSafe
Essentials

National Restaurant Association
EDUCATIONAL FOUNDATION

DISCLAIMER

The information presented in this book has been compiled from sources and documents believed to be reliable and represents the best professional judgment of the National Restaurant Association Educational Foundation. However, the accuracy of the information presented is not guaranteed, nor is any responsibility assumed or implied, by the National Restaurant Association Educational Foundation for any damage or loss resulting from inaccuracies or omissions.

Laws may vary greatly by city, county, or state. This book is not intended to provide legal advice or establish standards of reasonable behavior. Operators who develop food safety related policies and procedures as part of their commitment to employee and customer safety are urged to use the advice and guidance of legal counsel.

Table of Contents

International Food Safety Council

A Strategic Initiative of the
National Restaurant Association
EDUCATIONAL FOUNDATION

In 1993, the National Restaurant Association Educational Foundation recognized the need for food safety awareness and created the International Food Safety Council (Council). The Council's mission is to heighten the awareness of the importance of food safety education throughout the restaurant and foodservice industry.

The Council envisions a future where foodborne illness no longer exists. Through its educational programs, publications and awareness campaigns the Council reaches foodservice instructors, restaurant operators, suppliers, manufacturers, distributors, as well as organizations such as healthcare facilities, colleges and universities, supermarkets, associations and other stakeholders in the industry.

Founding Sponsors

American Egg Board
847.296.7043
www.aeb.org

Campbell Soup Company
800.879.7687
www.campbellsoup.com

ECOLAB, Inc.
800.352.5326
www.ecolab.com

FoodHandler Inc.
800.338.4433
www.foodhandler.com

Heinz North America
800.547.8924
www.heinz.com

KatchAll/San Jamar
800.533.6900
www.katchall.com
www.sanjamar.com

National Cattlemen's Beef Association
303.694.0305
www.beef.org
www.beeffoodservice.org

SYSCO Corporation
281.584.1390
www.sysco.com

Tyson Foods, Inc.
800.424.4253
www.tyson.com

Campaign Sponsors

Activ USA, Inc.
877.228.4879
www.activusa.net

Alaska Seafood Marketing Institute
800.806.2497
www.alaskaseafood.org

Atkins Temptec
800.284.2842
www.atkinstemptec.com

Bunzl Distribution
888.997.4515
www.bunzldistribution.com

Cargill Foods
800.CARGILL
www.cargillfoods.com

Colgate-Palmolive Company
888.276.0783
www.colpalipd.com

Cooper Instrument Corporation
800.835.5011
www.cooperinstrument.com

Daydots International
800.321.3687
www.daydots.com

Farquharson Enterprises, Ltd.
800.773.1455

International Dairy-Deli-Bakery Association™
608.238.7908
www.iddba.com

International Foodservice Manufacturers Association (IFMA)
312.540.4400
www.foodserviceworld.com/ifma

Johnson Wax Professional
800.558.2332
www.jwp.com

Jones Dairy Farm
800.635.6637
www.jonessausage.com

Lipton Foodservice
800.884.4841
www.liptonfs.com

Monsanto Company
314.694.1000
www.monsanto.com

Orkin Exterminating Company, Inc.
800.ORKIN.NOW
www.orkin.com

Procter & Gamble Company
800.817.6710
www.pg.com

Produce Marketing Association
302.738.7100
www.pma.com

Reckitt Benckiser, Inc.
800.677.9218
www.reckittprofessional.com

U.S. Foodservice, Inc.
410.312.7100
www.usfoodservice.com

For more information on the National Restaurant Association Educational Foundation's International Food Safety Council, visit our Web site at www.foodsafetycouncil.org or contact us at:
International Food Safety Council
250 South Wacker Drive, Suite 1400
Chicago, IL 60606
800.456.0111
312.715.1010 In Chicagoland

Acknowledgements

The development of the ServSafe Essentials would not have been possible without the expertise of our many advisors and manuscript reviewers. The National Restaurant Association Educational Foundation is pleased to thank the following people for the time and effort they dedicated to this project.

Marie-Luise Baehr, Sodexho Marriott Services

Cheryl Barsness, Alaska Seafood Marketing Institute

Jeff Carletti, KatchAll Industries

Elaine Cash, Daydots International

Larry Clark, Travel Centers of America

Gary DuBois, Taco Bell Corporation

Deborah Fitzgerald, Cooper Instrument Corporation

Kristen Forrestal, Darden Restaurants, Inc.

Peter Good, Peter Good Seminars, Inc.

David Goronkin, Buffets, Inc.

Joe Grosdidier, Ecolab, Inc.

Steven F. Grover, R.E.H.S., National Restaurant Association

Margaret Hardin, Ph.D, National Pork Producers Council

Harold Harlan, Ph.D., National Pest Control Association

Jean Hayden, Ohio Department of Health

Alice M. Heinze, RD, MBA, American Egg Board

Michael P. Hiza, RD, Manchester Community Technical College

Jane Lindeman, National Cattlemen's Beef Association

Mark McFarlane, Briazz

Kathy Means, Produce Marketing Association

Bruce Moilan, Chester-Jensen Co., Inc.

Nevin B. Montgomery, National Frozen Food Association, Inc.

Mauro Mordini, Lipton Foodservice

Robert Munnis, CSFP, Atkins Technical, Inc.

Alain Porte, MBA, SmithKline Beecham Pharmaceuticals

David J. Poulter, Buffets, Inc.

Mary Sandford, Burger King Corporation

Jill A. Snowdon, American Egg Board

J.R. (Jim) Starnes, FMP, Tyson Foods, Inc.

Lisa Wright, Foodmaker, Inc./Jack in the Box Restaurants

A Message From

THE NATIONAL RESTAURANT ASSOCIATION EDUCATIONAL FOUNDATION

Food safety is non-negotiable. Serving safe food is not an option. It's our obligation as restaurant and food service professionals. Proper training is one of the best ways to create a culture of food safety within our establishments.

By opening this book you have made a significant commitment to promoting food safety. We applaud you for that commitment.

The ServSafe® program has become the industry standard in food-safety training and is accepted in almost all United States jurisdictions that require employee certification. The ServSafe program provides accurate, up-to-date information for all levels of employees on all aspects of handling food, from receiving and storing to preparing and serving. You will learn science-based information on how to run a safe establishment— information all of your employees need to have in order to be a part of the food-safety team.

Your food-safety education does not end once you are certified in the ServSafe program. You have the responsibility to take your knowledge back to the unit and make your coworkers part of the food-safety culture. You will qualify to participate in the International Food Safety Council, which will help you share your knowledge. The Council, a coalition of restaurant and foodservice professionals created by the National Restaurant Association Educational Foundation in 1993, is a valuable resource that promotes the importance of food-safety training.

Whether you are new to the restaurant industry or continuing your career, your participation in ServSafe training will make you more qualified to serve safe food and to spread that knowledge throughout the industry.

Thank you for making the commitment to food-safety training and becoming an active part of the food-safety culture within the rapidly growing restaurant, foodservice, and hospitality industry.

Features of the ServSafe Essentials

We have designed the ServSafe Essentials to enhance your ability to learn and retain the food-safety knowledge that is essential to keep your establishment safe. Here are the key features you can expect to find in each section of this book.

TEST YOUR FOOD-SAFETY KNOWLEDGE:

Each section begins with five True or False questions, which are designed to test your prior knowledge of some of the concepts presented in the section. To find the answer to each question, and for further explanation, go to the heading and page number indicated.

LEARNING ESSENTIALS:

The learning essentials identify the things you should be able to do after completing each section. These objectives are linked to the tasks essential to keep your establishment safe.

CONCEPTS:

Concepts that are important to a thorough understanding of food safety are identified in the beginning of each section.

ARTWORK:

Photographs, illustrations, charts, and tables have been placed throughout each section, which visually reinforce or review key concepts presented in the text.

A CASE IN POINT:

These real-world scenarios give you the opportunity to apply concepts that you have learned in the section.

TRAINING TIPS FOR THE CLASSROOM:

These tips appear at the end of the section and are designed to give you training tools that you can use to teach others the food-safety concepts presented in the section. Each training tip begins with an objective, which tells you what your learner should be able to do after completing the activity. These tips are best suited for teaching food safety in a classroom environment.

TRAINING TIPS ON THE JOB:

These tips appear at the end of the section and are designed to help you implement food-safety concepts back at your establishment with your personnel. Specifically, these are designed to aid you in sharing the food-safety knowledge you have acquired. Each training tip begins with a statement that describes the purpose of the activity.

ACTIVITIES:

Various types of application activities appear right after the summary of the section. They provide you with the opportunity to practice and reinforce the concepts presented in the section and include crossword puzzles, word finds, finding the errors in illustrations, role-plays, and problem solving.

MULTIPLE-CHOICE STUDY QUESTIONS:

These questions are designed to test your knowledge of the food-safety concepts presented in the section. These questions are similar to the ones that you will find on the certification exam. If you have difficulty answering these questions, review the content further.

TAKE HOME IDEAS

TAKE HOME IDEAS

UNIT 1

FOOD SAFETY'S IMPACT ON THE OPERATION

 Food safety is an integral element of our organizational culture, which incorporates mutual respect and trust among guests, employees and vendors. Our guests expect not only a high quality dining experience, but a safe dining experience as well. The ServSafe Food Safety Training Program is the foundation we use to build awareness at all levels in the organization. Every manager in our system has been ServSafe certified. Commitment to food safety has the same organizational relevance as great tasting food, excellent service, and a clean environment.

David Goronkin
Executive Vice President, Operations
Buffets, Inc.

Section 1
Providing Safe Food

After completing this section, you should be able to:

○ Describe foodborne illness as a threat to the foodservice industry.

○ List populations that may be at high risk for foodborne illness and explain why.

○ Outline the characteristics of potentially hazardous foods.

○ Describe how foodborne illness happens.

○ Identify potential sources of food contaminants associated with human contact with food.

○ Recognize time-temperature abuse, proper personal hygiene, avoiding cross-contamination, and buying from approved sources as the key practices for ensuring food safety.

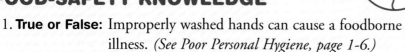

TEST YOUR FOOD-SAFETY KNOWLEDGE

1. **True or False:** Improperly washed hands can cause a foodborne illness. *(See Poor Personal Hygiene, page 1-6.)*

2. **True or False:** Young children may be more likely than adults to become ill from contaminated food. *(See Populations at High Risk for Foodborne Illness, page 1-4.)*

3. **True or False:** Food has been time-temperature abused any time it has been allowed to remain at temperatures favorable to the growth of microorganisms. *(See Time-Temperature Abuse, page 1-6.)*

4. **True or False:** A foodhandler's dirty clothing can cross-contaminate food. *(See Practicing Good Personal Hygiene, page 1-7.)*

5. **True or False:** Potentially hazardous foods are generally dry, low in protein, and highly acidic. *(See Foods Most Likely to Become Unsafe, page 1-4.)*

CONCEPTS

○ **Foodborne illness:** A disease that is carried or transmitted to people by food.

○ **Outbreak of foodborne illness:** An incident in which two or more people experience the same illness after eating the same food.

○ **Hazard Analysis Critical Control Point (HACCP):** A dynamic process that uses a combination of proper foodhandling procedures, monitoring techniques, and record keeping to help ensure food safety.

○ **Flow of food:** The path food takes from receiving and storage through preparation and cooking, holding, serving, cooling and reheating.

○ **The Model Food Code:** A science-based reference for retail restaurants and establishments on how to prevent foodborne illness.

○ **Contamination:** The presence of harmful substances not originally present in the food. Some food-safety hazards occur naturally while others are introduced by humans or the environment.

○ **Time-temperature abuse:** A food has been time-temperature abused any time it has been allowed to remain for too long at temperatures favorable to the growth of microorganisms.

○ **Potentially hazardous foods:** Foods in which microorganisms can rapidly grow. Potentially hazardous foods often have a history of being involved in foodborne outbreaks, have potential for contamination due to methods used to produce and process them, and have characteristics that generally allow microorganisms to thrive. They are often moist, high in protein, and chemically neutral or slightly acidic.

○ **Cross-contamination:** Occurs when microorganisms are transferred from one surface or food to another.

○ **Personal hygiene:** Sanitary health habits that include keeping body, hair, and teeth clean; wearing clean clothes; and washing hands regularly, especially when handling food and beverages.

INTRODUCTION

When diners eat out, they expect safe food, clean surroundings, and well-groomed workers. Overall, the restaurant and foodservice industry does a good job of meeting these demands, but there is still room for improvement.

The risk of foodborne illness impacts the industry. Several factors account for this, and likely include the following:

○ The emergence of new foodborne pathogens (disease-causing organisms)

○ The importation of food from countries where food-safety practices may not be well developed

○ Changes in the composition of foods, which may leave fewer natural barriers to the growth of microorganisms

○ Increases in the purchase of take-out and home meal replacement (HMR) foods

○ Changing demographics, with an increased number of individuals at high risk for contracting foodborne illness

○ Employee turnover rates that make it difficult to manage an effective food-safety system

THE DANGERS OF FOODBORNE ILLNESS

The greatest dangers to food safety are foodborne illnesses. A foodborne illness is a disease that is carried or transmitted to people by food. The Centers for Disease Control and Prevention (CDC) defines an outbreak of foodborne illness as an incident in which two or more people experience the same illness after eating the same food. A foodborne illness is confirmed when laboratory analysis shows that a specific food is the source of the illness.

Each year, millions of people become ill from foodborne illness, although the majority of cases are not reported and do not occur at restaurants or foodservice establishments. However, the cases that are reported and investigated help us understand some of the causes of illness, and what we, as restaurant and foodservice professionals, can do to control these causes in each of our establishments. The following are the most commonly reported causes of foodborne illnesses.

○ Failure to properly cool foods

○ Failure to cook and hold foods at the proper temperature

○ Poor personal hygiene

Fortunately, every restaurant and establishment, no matter how large or small, can take steps to ensure the safety of the food it prepares and serves to its customers.

The Costs of Foodborne Illness

National Restaurant Association figures show that a foodborne-illness outbreak can cost an establishment thousands of dollars. It can even cause an establishment to close. The illustration on the right outlines additional costs of a foodborne-illness outbreak.

Loss of Customers and Sales

Loss of Prestige and Reputation

Lawsuits Resulting in Lawyer and Court Fees

Increased Insurance Premiums

Lowered Employee Morale

Absenteeism of Employees

Need for Retraining Employees

Embarrassment

Cost of a Foodborne Illness to an Establishment

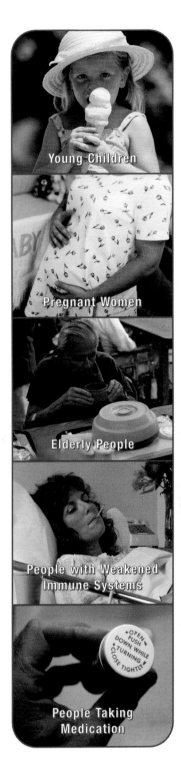

People at High Risk for Foodborne Illness

PREVENTING FOODBORNE ILLNESS

There are many challenges to preventing foodborne illness. These include the following:

○ High employee turnover rates

○ Service to an increasing number of high-risk customers

○ The service of potentially hazardous foods

Food-Safety Programs

An establishment should have an effective, proactive program that is based on preventing food-safety hazards before they occur. A reactive program that corrects a problem after it has occurred is not an effective system. The Hazard Analysis Critical Control Point (HACCP) program is a proactive, comprehensive, science-based food-safety system that allows operators to continuously monitor their establishments and reduce the risk of foodborne illness.

A HACCP system emphasizes the following:

○ How food flows through the operation from receiving to reheating

○ Identification of the points in the operation where contamination or growth of microorganisms can occur: control procedures can then be implemented based on the hazards identified at those points

Populations at High Risk for Foodborne Illness

The demographics of our population show there is an increase in the percentage of people at high risk of contracting a foodborne illness. These individuals are indicated to the left.

○ Young children are more at risk for contracting foodborne illnesses because they have not yet built up adequate immune systems (the body's defense system against illness) to deal with some diseases.

○ Elderly people are more at risk because their immune systems and resistance may have weakened with age.

Foods Most Likely to Become Unsafe

Although any food can become contaminated, most foodborne illnesses are transmitted through foods in which microorganisms are able to grow rapidly. Such foods are classified as potentially hazardous foods. These foods typically have the following characteristics.

○ A history of being involved in foodborne-illness outbreaks

○ A natural potential for contamination due to methods used to produce and process them

○ Moisture

○ High in protein

○ Neutral or slightly acidic pH

The FDA Model Food Code identifies the foods illustrated below as potentially hazardous foods.

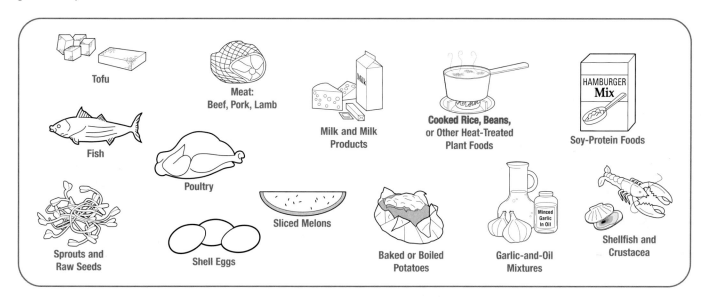

Potentially Hazardous Foods

Potential Hazards to Food Safety

Unsafe food usually results from contamination, which is the presence of harmful substances not originally present in the food. Some food-safety hazards are introduced by humans or by the environment, and some occur naturally.

Food-safety hazards are divided into three categories: biological hazards, chemical hazards, and physical hazards.

○ Biological hazards include certain bacteria, viruses, parasites, and fungi, as well as certain plants, mushrooms, and fish that carry harmful toxins.

○ Chemical hazards include pesticides, food additives, preservatives, cleaning supplies, and toxic metals that leach from cookware and equipment.

○ Physical hazards consist of foreign objects that accidentally get into the food, such as hair, dirt, metal staples, and broken glass.

By far, biological hazards pose the greatest threat to food safety. Disease-causing microorganisms are responsible for the majority of foodborne illness outbreaks.

HOW FOOD BECOMES UNSAFE

Foodborne illness is caused by several factors, which can be placed into one of three categories: time-temperature abuse, cross-contamination, and poor personal hygiene. Reported cases of foodborne illness usually involve more than one factor in each of these categories. A well-designed food-safety system will control these factors.

Time-Temperature Abuse: Food has been time-temperature abused any time it has been allowed to remain for too long at temperatures favorable to the growth of microorganisms. Common factors that have resulted in foodborne illness include the following.

○ Failure to hold or store food at required temperatures

○ Failure to cook or reheat foods to temperatures that kill microorganisms

○ Failure to properly cool foods

○ Preparation of foods a day or more before they are served

Cross-Contamination: Cross-contamination occurs when microorganisms are transferred from one surface or food to another. Common factors that have resulted in foodborne illness include the following:

○ Adding raw, contaminated ingredients to foods that receive no further cooking

○ Food-contact surfaces (such as equipment or utensils) that are not cleaned and sanitized before touching cooked or ready-to-eat foods

○ Allowing raw food to touch or drip fluids onto cooked or ready-to-eat food

○ Hands that touch contaminated (usually raw) food and then touch cooked or ready-to-eat food

○ Contaminated cleaning cloths that are not cleaned and sanitized before being used on other food-contact surfaces

Poor Personal Hygiene: Individuals with unacceptable personal hygiene can offend customers, contaminate food or food-contact surfaces, and cause illnesses. Common factors that have resulted in foodborne illness include the following:

○ Employees who fail to properly wash their hands after using the restroom, or whenever necessary

○ Employees who cough or sneeze on food

○ Employees who touch or scratch sores, cuts, or boils and then touch food they are preparing or serving

KEY PRACTICES FOR ENSURING FOOD SAFETY

The key to food safety lies in controlling time and temperature throughout the flow of food, practicing good personal hygiene, preventing cross-contamination, and purchasing food supplies from approved suppliers. It is important to establish standard operating procedures that focus on these areas.

Controlling Time and Temperature

Microorganisms pose the largest threat to food safety. Like all living organisms, they cannot survive or reproduce outside certain temperature limits. The chart to the right shows how time and temperature is controlled throughout the flow of food to minimize microbial growth.

Practicing Good Personal Hygiene

Training employees in good personal hygiene is the responsibility of every manager. Features of a good personal hygiene program include the following:

○ **Proper handwashing.** Hands and fingernails should be washed and cleaned thoroughly before handling food, between each task, and before using food-preparation equipment.

○ **Strictly enforced rules regarding eating, drinking, and smoking.** These activities should be prohibited while preparing or serving food, or while in areas used for washing equipment and utensils.

○ **Preventing employees who are ill from working with food.** Cuts, burns, and sores must be properly cleaned and covered.

○ **General cleanliness.** Insist on daily bathing, clean hair, and clean clothing.

Preventing Cross-Contamination

Employees must be carefully trained to recognize and prevent cross-contamination of microorganisms between foods and food-contact surfaces. Some ways to prevent cross-contamination include the following:

○ Require employees to wash their hands frequently when working with raw foods. They should never touch raw foods and then touch ready-to-eat foods without washing their hands.

○ Do not allow raw or contaminated food to touch or drip fluids onto cooked or ready-to-eat foods.

○ Clean and sanitize food-contact surfaces (such as equipment or utensils) that touch contaminated food before they come in contact with cooked or ready-to-eat foods.

○ Clean and sanitize cleaning cloths between each use.

Purchasing from Approved Suppliers

Use reputable and reliable suppliers to help avoid receiving contaminated foods. Suppliers' products and practices should meet your specifications as well as local, state, and federal regulations. Work closely with suppliers to set up effective receiving procedures, and make sure you address these in your purchase specification agreements. Reputable suppliers and distributors will address the following:

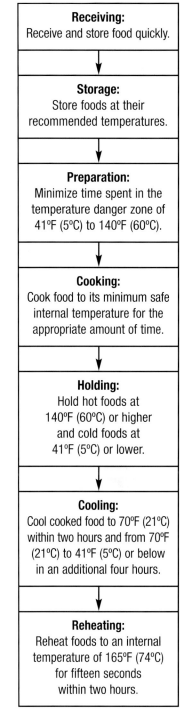

Receiving:
Receive and store food quickly.

Storage:
Store foods at their recommended temperatures.

Preparation:
Minimize time spent in the temperature danger zone of 41°F (5°C) to 140°F (60°C).

Cooking:
Cook food to its minimum safe internal temperature for the appropriate amount of time.

Holding:
Hold hot foods at 140°F (60°C) or higher and cold foods at 41°F (5°C) or lower.

Cooling:
Cool cooked food to 70°F (21°C) within two hours and from 70°F (21°C) to 41°F (5°C) or below in an additional four hours.

Reheating:
Reheat foods to an internal temperature of 165°F (74°C) for fifteen seconds within two hours.

Controlling Time and Temperature Throughout the Flow of Food

○ Deliver food at proper temperatures.

○ Use clean, and where appropriate, refrigerated trucks.

○ Train their employees in food-safety practices.

○ Use protective, leak-proof, durable packaging.

○ Agree to adjust delivery schedules to meet your establishment's needs.

○ Cooperate with your employees who inspect products when they are delivered.

○ Allow you to inspect their delivery vehicles and production facilities.

○ Make their inspection reports available, if asked.

SUMMARY

Foodborne illness is a major concern to the restaurant and foodservice industry. A foodborne illness is a disease that is carried or transmitted to people by food. An outbreak of foodborne illness usually involves two or more people experiencing the same illness after eating the same food. An incident of foodborne illness can be very expensive for an establishment, considering legal liability, damage to reputation, and other related factors. However, a well-designed food-safety program protects your customers, your employees, and your reputation. The Hazard Analysis Critical Control Point (HACCP) program helps you monitor operations and reduce the opportunities for foodborne illness. The HACCP system focuses on the flow of food in your operation, identifying points where contamination can occur, and implementing controls for those points. Some categories of people are more susceptible to foodborne illnesses than others. These categories are called high-risk populations. Some foods have a history of involvement in foodborne illness outbreaks. These are called potentially hazardous foods. Typically, they have a natural potential for contamination due to the methods used to produce or process them, and are often moist, high in protein, and have a neutral or slightly acidic pH. Food can become unsafe at any step in the flow of food. Control time and temperature and avoid contamination at each step. Work with reputable suppliers and implement strict receiving procedures to help ensure the delivery of safe food. Once the food arrives, it must be stored, prepared, cooked, held, served, cooled, and reheated using methods that maintain its safety. People pose a major risk to safe food, especially foodhandlers who do not practice personal hygiene. You must carefully train, monitor, and reinforce food-safety principles in your establishment. Establishing a well-designed food-safety system can help protect your customers by preventing outbreaks of foodborne illness and can help the establishment avoid the potentially high costs associated with them.

ACTIVITY

Crossword Puzzle

Across:

1. This is the path food takes from receiving and storage through preparation and cooking, holding, serving, cooling and reheating.

2. A disease that is carried or transmitted to people by food.

4. This is a science-based reference for retail restaurants and establishments on how to prevent foodborne illness.

8. They are often foods that are moist, high in protein, and chemically neutral or slightly acidic.

Down:

1. An incident in which two or more people experience the same illness after eating the same food.

3. Any time a food has been allowed to remain for too long at temperatures favorable to the growth of microorganisms.

5. Occurs when microorganisms are transferred from one surface or food to another.

6. Sanitary health habits that include keeping body, hair, and teeth clean; wearing clean clothes; and washing hands regularly, especially when handling food and beverages.

7. This is the presence of harmful substances not originally present in the food.

9. A dynamic process that uses a combination of proper foodhandling procedures, monitoring techniques, and record keeping to help ensure food safety.

ACTIVITY

Word Find

Find the terms that go with the clues below.

Clues

1. This is the presence of harmful substances not originally present in the food.

2. This is the path food takes from receiving and storage through preparation and cooking, holding, serving, cooling and reheating.

3. A disease that is carried or transmitted to people by food.

4. An incident in which two or more people experience the same illness after eating the same food.

5. This is a science-based reference for retail restaurants and establishments on how to prevent foodborne illness.

6. Occurs when microorganisms are transferred from one surface or food to another.

```
P F Z L W W N O I T A N I M A T N O C S S O R C H
Q O L V T U J X B B E D O C D O O F L E D O M A K
Y P T O X T I A Q W L E C Q A G B A Z I N I C A C
G E S E W I F D Q E I H L N B J V O A H X C E P O
H R F K N O J M L L O X B H M V J K O E P R Z K N
E S M Y O T F X X L L B Q W T H F W V G B V F A T
S O B F U V I F T N W Q R R V K B A A T L S I V A
U N P X L F H A O T T G Z I N F G S U R T R O S M
B A K W T L Y B L O A R X A M V Y O C K H N K U I
A L J P L B V L L L D H O B L Q S U G V U B K K N
E H Y T W B J F T H Y B U A F S Y W C A H Y T Z A
R Y W V A G P V G M R H Q K E Z N C H Q P H T M T
U G H U Q A C Q F T B N A N B N K T Z J H N Z V I
T I L R D A D T V R L K L Z X A H B H B H F G K O
A E V S L V X N S N A L T W A O W O F L W H N O N
R N S A X R A J Q A I S V N B R E V Y N I O M Z S
E E R T R J L F I E K J C G N Q D Z A W U J M G R
P D R Y A D T Z N O L O J T D K K O Q M D T X Q E
M B D G E A A R L V W L P H V Q X G U R P T C W K
E P Y P D O O J L G N B X Q E L E V V S N B I Y W
T V E D Q B S H M B O R A W A V D Y D I F L C B B
E Z A H D B E J E A V N O U A A N O W G F O Q G Y
M R R O L F C I S F I K J B N G V S P C C D O N U
I I O S T B D S C U C W A S M R N V U I Q B G D K
T F O W L D Y F S S E N L L I E N R O B D O O F S
```

7. A dynamic process that uses a combination of proper foodhandling procedures, monitoring techniques, and record keeping to help ensure food safety.

8. They are often foods that are moist, high in protein, and chemically neutral or slightly acidic.

9. Sanitary health habits that include keeping body, hair, and teeth clean; wearing clean clothes; and washing hands regularly, especially when handling food and beverages.

10. Any time a food has been allowed to remain for too long at temperatures favorable to the growth of microorganisms.

ACTIVITY

Potentially Hazardous or Not? That is the Question

Listed below are several food items. Circle the items that would be considered potentially hazardous foods based on the criteria explained in this section.

1. Bananas

2. Shell eggs

3. Bean sprouts

4. Baked potatoes

5. Soda crackers

6. Lettuce

7. Dry egg noodles

8. Sliced melons

9. Candy

10. Flour

11. Ground beef

12. Pickles

13. Limes

14. Fish

15. Cooked rice

16. Poultry

17. Vinegar

18. Jelly

ACTIVITY

Good and Bad Practices

This exercise can be done by individuals working alone or in small groups. Spend the first half of the available time preparing the guidelines, and the second half discussing them as a group.

In this section, we discussed the key practices necessary for ensuring safe food. Think about your establishment as it relates to these key practices. Identify all the things that you are doing right and all the things that you are doing wrong. For the things that you are doing wrong, develop guidelines to correct these practices.

TRAINING TIPS

Training Tips for the Classroom

1. Food Safety on Trial: "But your honor, I thought *they were washing their hands.*"

Objective: *After completing this activity, class participants will be able to identify the legal liability faced by establishments.*

Directions: Present your class with a case study of an outbreak of foodborne illness. Use one of the Case In Point examples from this text or create a detailed fictitious case. Have the class assume that the outbreak has resulted in several lawsuits.

Break the class into the following teams: a plaintiff team (representing the persons who claim they became ill), a defense team (representing the establishment that served the food), and a jury. Allow about ten minutes for the plaintiff and defense to plan their presentations, then another ten minutes for each side to present its argument before the judge (the instructor). Questions to be argued include the following.

- Was the food unfit to serve?
- Did the food cause the plaintiff harm?
- Did the restaurant or establishment violate the "warranty of sale"?
- Did the restaurant or establishment exhibit "reasonable care"?

After both sides have been presented, give the jury five minutes to deliberate and determine the fate of the restaurant or establishment. No mistrials or appeals!

2. Knowledge and Application of Food Safety

Objective: *After completing this activity, class participants should be able to identify information that every manager should know and apply at his or her establishment, as recommended by the FDA.*

Directions: The FDA recommends that local and state regulatory agencies hold the person in charge of a restaurant or establishment responsible for knowing and applying the information listed below. Assign teams of two or three people. Depending upon the number of teams, assign each team one or more topics from the list. Ask the teams to prepare a discussion of current food-safety knowledge related to their assigned topics. After the teams have had adequate time to prepare, ask them to present their information to the class. After the presentation, solicit group discussion. You might ask class members how well the management teams in their own establishments follow the FDA recommendations.

The FDA recommends that the person in charge of a restaurant or establishment know and apply the following information.

○ The diseases that are carried or transmitted by food and the symptoms of these diseases

○ Points in the flow of food where hazards can be prevented, eliminated, or reduced, and how procedures meet the requirements of the local code

○ The relationship between personal hygiene and the spread of disease, especially concerning cross-contamination, hand contact with ready-to-eat foods, and handwashing

○ How to keep injured or ill employees from contaminating food or food-contact surfaces

○ The need to control the length of time that potentially hazardous foods remain at temperatures where disease-causing microorganisms can grow

○ The hazards involved in the consumption of raw or undercooked meat, poultry, eggs, and fish

○ Safe cooking temperatures and times for potentially hazardous foods, such as meat, poultry, eggs, and fish

○ Safe temperatures and times for the safe refrigerated storage, hot holding, cooling, and reheating of potentially hazardous foods

○ Correct procedures for cleaning and sanitizing utensils and food-contact surfaces of equipment

○ The types of poisonous and toxic materials used in the operation, and how to safely store, dispense, use, and dispose of them

○ The need for equipment that is sufficient in number and capacity, and is properly designed, constructed, located, installed, operated, maintained, and cleaned

○ The sources of the operation's water supply and the importance of keeping it clean and safe

○ The rights, responsibilities, and authorities the local code assigns to employees, managers, and the local health department

Training Tips on the Job

1. Developing a Food-Safety Complaint Form for Your Establishment

Purpose: *To involve your management team in the development of a standard complaint questionnaire form for your establishment. Note: Involving the management team will help clarify your policies regarding how to handle a possible complaint of foodborne illness.*

Directions: Every establishment should have a standard complaint questionnaire form on hand to record customer reports of possible foodborne illness. These forms are essential in order to ensure that complaints are handled in a professional manner, information is fully documented, and all appropriate questions are included.

Ask members of the management team to help write and review questions for the form. It is a good idea to look at similar forms from other establishments. Ask your local regulatory agency or state restaurant association for input. If possible, you may want to have a lawyer review the form. When the content of the form is approved, print up copies and make them accessible to the management team.

Inform your team that it is essential to use these forms whenever a customer has a complaint about foodborne illness. Let the team know how you want a complaint to be handled. They need to know what they should say to the customer, and what steps to take once the form has been filled out. It may be a good idea to role-play scenarios regarding a customer complaint. This will allow managers to practice, and can expose any weaknesses that may exist in your program.

2. Marketing Your Safe Food

Purpose: *To involve your management team in a brainstorming session to determine different ways to market the food safety of your establishment to both employees and customers.*

Directions: Conduct a brainstorming session with your management team to determine different ways to market your food-safety systems and procedures to employees and customers. If you already market your systems and procedures, use the brainstorming session to improve your current marketing practices.

Bring up the following questions in the session:

Employees

○ How do we demonstrate to our employees that we are committed to food safety?

○ What food-safety systems are currently in place? Do foodhandlers understand these systems?

○ What training commitment have we made regarding food safety? How is this training made available to employees?

○ What documentation do we keep related to food safety? Do we monitor Critical Control Points? If so, do employees understand how to accurately monitor and document Critical Control Points?

Customers

○ How is our commitment to food safety visible to our customers?

○ What food-safety principles do our customers see being put into practice by our staff?

○ What written materials (signs, plaques, posters, decals, statements on packaging, etc.) are in place to inform our customers of our commitment to food safety?

Based on the input received from your management team, come up with a plan to market your food-safety systems and procedures to employees and customers, or to improve your current marketing practices. Implement the plan and evaluate it at management meetings. If necessary, modify your plan based on the evaluation.

MULTIPLE-CHOICE STUDY QUESTIONS

1. Why do people who live with chronic illnesses have a higher risk of contracting a foodborne illness?

   A. They are likely to eat food prepared in large quantities while they are in a hospital.

   B. Their internal disease-fighting systems are likely to be weaker than normal.

   C. Their allergic reactions to chemicals used in food production might be greater than normal.

   D. They are likely to have diminished appetites and do not want to cook for themselves.

2. Which type of food would be the most likely to cause a foodborne illness?

   A. Tomato juice
   B. Baked potatoes
   C. Stored whole wheat flour
   D. Dry powdered milk

3. Your restaurant is closed Sunday and Monday. Tuesday morning you open the restaurant and notice that the refrigerator is not running. When you check the internal thermometer, it reads 50°F (10°C). What should you do with the fresh ground beef in the refrigerator?

   A. Cook and serve it within two hours.
   B. Freeze it right away.
   C. Discard it.
   D. Re-chill it immediately to below 41°F (5°C).

4. Which of the following is not a common characteristic of potentially hazardous foods?

   A. They are moist.
   B. They are dry.
   C. They are neutral or slightly acidic.
   D. They are high in protein.

5. When a foodborne illness occurs, it is usually caused by a combination of three factors. These factors are
   A. time-temperature abuse, unacceptable personal hygiene, and physical hazards.
   B. time-temperature abuse, chemical hazards, and physical hazards.
   C. time-temperature abuse, cross-contamination, and chemical hazards.
   D. time-temperature abuse, cross-contamination, and unacceptable personal hygiene.

6. In order for a foodborne illness to be considered an "outbreak," how many people must experience the illness after eating the same food?
   A. 1   B. 2   C. 10   D. 20

7. An effective food-safety system, such as HACCP, will
   A. react to food-safety hazards quickly after they occur.
   B. prevent food-safety hazards before they occur.
   C. be based on standards set up by establishments like yours.
   D. be based on standards mandated by the government.

8. The largest threat to food safety comes from which of the following contaminants?
   A. Pesticides   C. Microorganisms
   B. Hair   D. Food additives

9. To prevent cross-contamination, foodhandlers should not
   A. touch raw meat and then touch cooked or ready-to-eat food.
   B. allow food to remain at temperatures above 41°F (5°C).
   C. take food temperatures when receiving food.
   D. hold food at temperatures below 140°F (60°C).

10. Foodhandlers must practice all of the following hygienic practices except
    A. proper handwashing.
    B. daily bathing.
    C. wearing clean clothing to work.
    D. getting periodic AIDS tests.

Section 2
The Microworld

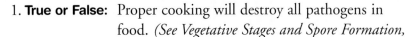

Knowledge

TEST YOUR
FOOD-SAFETY KNOWLEDGE

1. **True or False:** Proper cooking will destroy all pathogens in food. *(See Vegetative Stages and Spore Formation, page 2-5.)*

2. **True or False:** A foodborne intoxication is caused by eating food which contains a pathogen. *(See Foodborne Infection versus Foodborne Intoxication, page 2-17.)*

3. **True or False:** Most foodborne illnesses are caused by viruses. *(See Bacteria, page 2-4.)*

4. **True or False:** Bacteria will not grow at refrigeration temperatures. *(See Temperature, page 2-7.)*

5. **True or False:** Highly acidic foods such as tomato sauce provide a perfect environment for bacteria to grow in. *(See Acidity, page 2-7.)*

Learning Essentials

After completing this section, you should be able to:

○ Identify the four basic types of microbial contaminants, give examples, and describe preventive actions for each.

○ Differentiate between foodborne infections and foodborne intoxication and identify the major causes of each.

○ Identify the microbial risks associated with various types of food.

○ Explain the conditions conducive to bacterial growth.

CONCEPTS

○ **Microorganisms:** Small, living organisms that can be seen only with the aid of a microscope. The four types of microorganisms that can contaminate food and cause foodborne illness are bacteria, viruses, parasites, and fungi.

○ **Pathogens:** Disease-causing microorganisms.

○ **Bacteria:** Single-celled living microorganisms that can cause food spoilage and disease. Some form spores which can survive freezing and very high temperatures. They are more commonly involved in foodborne illness than are viruses, fungi or parasites.

○ **Virus:** The smallest of the microbial food contaminants. They rely on a living host to reproduce. Viruses usually contaminate food through a foodhandler's improper personal hygiene. Some may survive freezing and cooking temperatures.

○ **Parasite:** An organism that needs to live in or on a host organism to survive. Parasites can live inside many animals that humans use for food, such as cows, chickens, hogs, and fish. Proper cooking and freezing will kill parasites. Avoiding cross-contamination and proper handwashing can also prevent foodborne illness caused by parasites.

○ **Fungi:** Fungi range in size from microscopic, single-celled organisms to very large, multi-cellular organisms. Molds, yeasts and mushrooms are examples of fungi.

○ **pH:** A measure of a food's acidity or alkalinity. The pH scale ranges from 0 to 14.0. A pH above 7.0 is alkaline, while a pH below 7.0 is acidic. A pH of 7.0 is neutral. Pathogenic bacteria grow well in food with a pH between 4.6 and 7.5.

○ **Spore:** An alternative form for some bacteria that protects them from adverse conditions. The spore's thick walls protect the bacteria from high and low temperatures, low moisture, and high acidity. The spore is capable of turning back into a vegetative cell when conditions become favorable again.

○ **Vegetative microorganism:** Those bacteria that are in the process of reproduction. Bacteria reproduce by splitting in two. As long as conditions are favorable, bacteria can grow and multiply very rapidly, doubling their number as often as every twenty minutes.

○ **FAT-TOM:** An acronym for the conditions needed by most microorganisms to grow (except viruses): Food, Acidity, Time, Temperature, Oxygen, Moisture.

○ **Temperature Danger Zone:** The temperature range between 41°F and 140°F (5°C and 60°C) within which most bacteria grow and reproduce.

○ **Water activity:** The amount of moisture in food available for the growth of microorganisms. Potentially hazardous foods have water-activity values of 0.85 or above.

○ **Mold:** A type of fungus that causes food spoilage. Some molds produce toxins that can cause foodborne illness.

○ **Yeast:** A type of fungus that causes food spoilage.

○ **Foodborne infection:** An infection that results when pathogens grow in the intestines after a person eats food contaminated by them.

○ **Foodborne intoxication:** An intoxication caused by eating food containing poisonous toxins.

○ **Foodborne toxin-mediated infection:** An illness that occurs from eating food that contains pathogens. These pathogens grow in the intestines and produce toxins that can make you ill.

INTRODUCTION

In the previous section, you learned that microorganisms pose the greatest threat to food safety, and that disease-causing microorganisms are responsible for the majority of foodborne-illness outbreaks. In this section, you will learn about the microorganisms that cause foodborne illness, as well as the conditions they require in order to grow. When you understand these conditions, you will begin to see how the growth of microorganisms can be controlled, a topic that will be covered in greater detail in later chapters.

Microorganisms are small, living beings that can only be seen with a microscope. While not all microorganisms cause disease, some do. These are called pathogens. Eating food contaminated with pathogens or their toxins (poisons) is the leading cause of foodborne illness.

MICROBIAL CONTAMINANTS

There are four types of microorganisms that can contaminate food and cause foodborne illness: bacteria, viruses, parasites, and fungi.

These microorganisms can be arranged into two groups: spoilage microorganisms and pathogens. Mold is a spoilage microorganism. While its appearance, smell, and taste is not very appetizing, it typically does not cause illness. Pathogens like Salmonella, the virus that causes Hepatitis A, and *E. coli* O157:H7 cause some form of illness when ingested. However, unlike spoilage microorganisms, pathogens cannot be seen, smelled, or tasted.

BACTERIA

Bacteria are more commonly involved in foodborne illness than any of the other foodborne microorganisms. Knowing what bacteria are, and understanding the environment in which they grow, is the first step in controlling them.

Basic Characteristics of Bacteria that Cause Foodborne Illness

Bacteria that cause foodborne illness have some basic characteristics.

○ They are living, single-celled organisms.

○ They may be carried by a variety of means: food, water, humans, and insects.

○ Under favorable conditions, they can reproduce very rapidly.

○ Some can survive freezing.

○ Some form into spores, a change that protects the bacteria from unfavorable conditions.

○ Some can cause food spoilage while others can cause disease.

○ Some cause illness by producing toxins as they multiply, die, and break down.

Bacterial Growth

To grow and reproduce, bacteria need the following:

○ Adequate time

○ Proper temperature

○ Ample moisture

○ Food

○ Appropriate pH (acidity)

○ The necessary level of oxygen

Their growth can be broken down into four progressive stages (phases): lag, log, stationary, and death.

Lag Phase

○ The lag phase is the adjustment period for bacteria when it is first introduced to a food.

○ The bacteria's numbers are stable and they are preparing for growth.

○ Keep the bacteria in the lag phase as long as possible to control its growth.

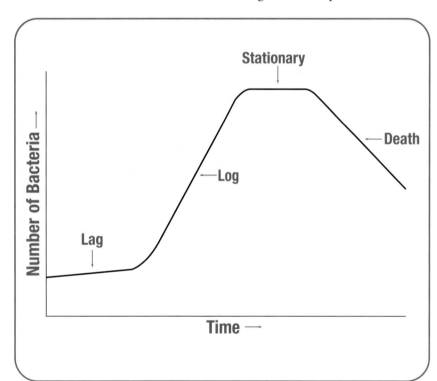

Growth Stages of Bacteria

This is accomplished by controlling the requirements for growth: time, temperature, moisture, oxygen, and pH.

○ If these conditions for growth are not controlled, bacteria will grow remarkably fast.

Log Phase

○ Bacteria reproduce by splitting in two. Those that are in the process of reproduction are called vegetative microorganisms.

○ As long as conditions are favorable, bacteria can grow and multiply very rapidly, doubling their number as often as every twenty minutes, as illustrated below.

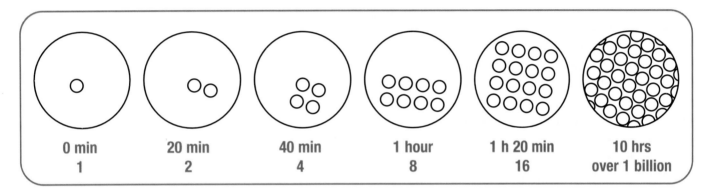

0 min	20 min	40 min	1 hour	1 h 20 min	10 hrs
1	2	4	8	16	over 1 billion

Rapid Bacterial Growth

Stationary Phase

○ Bacteria can continue to grow until nutrients and moisture become scarce, or conditions become unfavorable, at which point they begin to die.

○ Bacteria begin to die when resources become scarce. Eventually, the population reaches a stationary phase, where just as many bacteria are growing as are dying.

Death Phase

○ When the number of bacteria dying exceeds the number that are growing, the population declines.

Vegetative Stages and Spore Formation

Although vegetative bacteria may be resistant to low and even freezing temperatures, they can be killed by high temperatures. Some types of bacteria, however, have the ability to change into a different form, called a spore. The spore's thick wall protects the bacteria against unfavorable conditions, such as high or low temperature, low moisture, and high acidity.

While a spore cannot reproduce, it is capable of turning back into a vegetative organism when conditions become favorable again. For example, bacteria in food may form a spore when exposed to freezer temperatures, allowing the bacteria to survive. As the food thaws and conditions change, the spore can turn back into a vegetative cell and begin to grow in the food.

Since spores are so difficult to destroy, it is important to cook, cool, and reheat foods properly in order to keep bacteria that may be in food from growing to harmful levels.

FAT-TOM: What Microorganisms Need to Grow

The conditions that favor the growth of most foodborne microorganisms (except viruses) can be remembered by the acronym FAT-TOM. Each of these conditions for growth will be explained in more detail in the next several paragraphs.

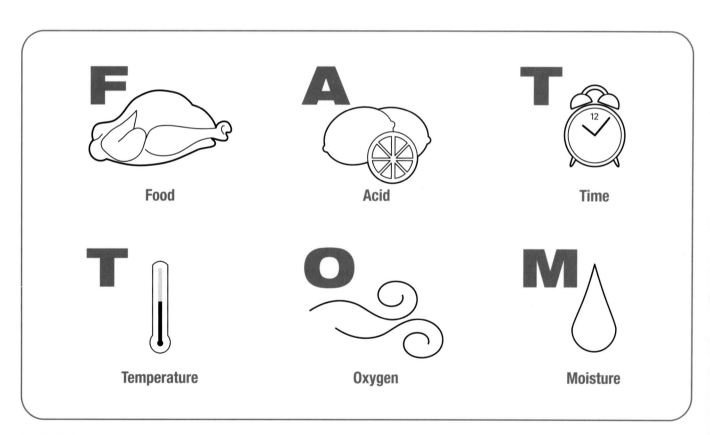

F Food   **A** Acid   **T** Time

T Temperature   **O** Oxygen   **M** Moisture

FAT-TOM
The conditions that favor the growth of most foodborne microorganisms (except viruses) can be remembered by the acronym FAT-TOM.

Food

○ To grow, microorganisms need nutrients, specifically proteins and carbohydrates. These proteins are commonly found in foods such as meat, poultry, dairy products, and eggs.

 Acidity

○ Microorganisms typically do not grow in foods that are highly acidic or highly alkaline.
○ Pathogenic bacteria grow well in foods with a pH between 4.6 and 7.5.
○ Foods with a pH higher than 7.0 do not typically support the growth of foodborne microorganisms.

 Temperature

○ Most foodborne microorganisms grow well between the temperatures of 41°F and 140°F (5°C to 60°C). This range is known as the temperature danger zone.
○ Exposing microorganisms to temperatures outside the danger zone does not necessarily kill them. Refrigeration temperatures, for example, may only slow their growth. Some bacteria grow well at refrigeration temperatures. Bacterial spores can often survive extreme heat and cold.

 Time

○ Microorganisms need sufficient time to grow. Bacteria can double their population every twenty minutes.
○ If contaminated food remains in the temperature danger zone for four hours or more, pathogenic microorganisms can grow to levels high enough to make someone ill.

 Oxygen

○ Most microorganisms that cause foodborne illness can grow with (aerobic) or without (anaerobic) the presence of oxygen.

 Moisture

○ Most foodborne microorganisms grow well in moist foods.
○ The amount of moisture in a food is called its water activity (a_w). Water activity is measured on a scale from 0 through 1.0, with distilled water having a water activity of 1.0.
○ Potentially hazardous foods typically have a water activity of 0.85 or above.

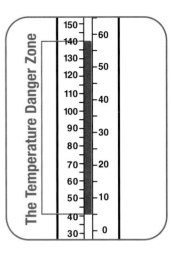

Temperature and Bacterial Growth
Most foodborne microorganisms grow well at temperatures between 41°F and 140°F (5°C and 60°C).

Major Foodborne Illnesses Caused by Bacteria

Foodborne Illness	Salmonellosis (nontyphoid)	Shigellosis	Listeriosis
Bacteria	*Salmonella* spp.	*Shigella* spp.	*Listeria monocytogenes*
Characteristics of Bacteria	Does not form spores; facultative; some can survive pHs below 4.5	Does not form spores; facultative; some produce shiga toxin	Does not form spores; facultative; resists freezing, drying, and heat; can grow at refrigeration temperatures
Type of Illness	Infection (possibly toxin-mediated)	Toxin-mediated infection	Infection
Symptoms	Abdominal cramps, headache, nausea, fever, diarrhea, and sometimes vomiting; may cause severe dehydration in infants and elderly; may cause arthritic symptoms 3 to 4 weeks later	Diarrhea (may be bloody), abdominal pain, fever, nausea, cramps, vomiting, chills, fatigue, dehydration	Nausea, vomiting, diarrhea, headache, persistent fever, chills, backache, meningitis, encephalitis, septicemia, and intrauterine or cervical infections in pregnant women, which may result in spontaneous abortion or stillbirth; most often affects fetuses, infants, and pregnant women
Incubation Period	6 to 48 hours; usually 12 to 36 hours	12 to 50 hours; usually 1 to 3 days	3 to 70 days; usually about 3 weeks
Duration of Illness	1 to 2 days (sometimes longer)	Usually 4 to 7 days; can be indefinite (depends on treatment)	Indefinite (depends on treatment); high fatality rates in immunocompromised people
Source	Water, soil, insects, domestic and wild animals, and the human intestinal tract, especially as carriers	Human intestinal tract; flies; frequently found in water polluted by feces	Soil, water, and damp environments; humans, domestic and wild animals, especially fowl (intestinal tracts)
Foods Involved in Outbreaks	Poultry and poultry salads; meat and meat products; fish; shrimp; milk; shell eggs and egg products, such as custards, sauces, and pastry creams; tofu and other protein foods; sliced melons, sliced tomatoes, raw sprouts, and other fresh produce	Salads (potato, tuna, shrimp, chicken, and macaroni); lettuce; raw vegetables, milk and dairy products; poultry; moist and mixed foods	Unpasteurized milk and cheese; ice cream; frozen yogurt; raw vegetables; poultry and meats; seafood; prepared and chilled ready-to-eat foods (e.g., soft cheese, deli, foods, pâté)
Preventive Measures	Avoid cross-contamination; refrigerate food; thoroughly cook poultry to at least 165°F (74°C) for at least 15 seconds and cook other foods to minimum internal temperatures; properly cool cooked meats and meat products within 6 hours; avoid pooling eggs; ensure that employees avoid contaminating food and food-contact surfaces by practicing good personal hygiene	Avoid cross-contamination; ensure that food-service employees practice good personal hygiene; use sanitary food and water sources; control flies; rapidly cool foods	Use only pasteurized milk and dairy products; cook foods to proper internal temperatures; avoid cross-contamination; clean and sanitize surfaces, keep facilities clean and dry

Major Foodborne Illnesses Caused by Bacteria (Continued)

Foodborne Illness	Staphylococcal Food Poisoning (Enterotoxicosis)	Clostridium perfringens Enteritis	Bacillus cereus Gastroenteritis
Bacteria	Staphylococcus aureus	Clostridium perfringens	Bacillus cereus
Characteristics of Bacteria	Does not form spores; facultative; can survive high acidity (to pH 2.6); resists drying and freezing	Forms spores; anaerobic	Forms spores; facultative
Type of Illness	Intoxication	Toxin-mediated infection	Intoxication (emetic); toxin-mediated infection (diarrheal)
Symptoms	Nausea, retching, abdominal cramps, diarrhea; in severe cases—headache, muscle cramping, changes in blood pressure and pulse rate	Abdominal pain, diarrhea, nausea, dehydration (fever, headache, and vomiting usually absent)	Nausea and vomiting, sometimes abdominal cramps or diarrhea (emetic); watery diarrhea, abdominal cramps, pain, nausea (diarrheal)
Incubation Period	1 to 7 hours; usually 2 to 4 hours	8 to 22 hours; usually 10 to 12 hours	30 minutes to 6 hours (emetic); 6 to 15 hours (diarrheal)
Duration of Illness	1 to 2 days	Usually 24 hours; may last 1 to 2 weeks	Less than 24 hours (emetic); 24 hours (diarrheal)
Source	Skin, hair, nose; throat; infected sores; animals	Humans and animals (intestinal tracts), soil; soil contaminated with feces	Soil and dust; cereal crops
Foods Involved in Outbreaks	Reheated foods; ham and other meats; poultry; egg products and other protein foods; milk and dairy products; potato sandwiches; custards; cream-filled pastries; salad dressings	Cooked meat; meat products; poultry; stew; gravy; beans that have been improperly cooled	Rice products; starchy foods (potato, pasta and cheese products); sauces; puddings; soups; casseroles; pastries; salads (emetic); meats; milk; vegetables; and fish (diarrheal)
Preventive Measures	Avoid contamination from unwashed bare hands; practice good personal hygiene; exclude foodservice employees with skin infections from foodhandling and preparation tasks; properly refrigerate food; rapidly cool prepared foods	Use careful time and temperature control in cooling and reheating cooked foods	Careful time and temperature control and quick-chilling methods to cool foods; adequate cooking of food

Major Foodborne Illnesses Caused by Bacteria (Continued)

Foodborne Illness	Botulism	Campylobacteriosis	E. coli 0157:H7 Enterohemorrhagic (EHEC)*
Bacteria	Clostridium botulinum	Campylobacter jejuni	Escherichia coli
Characteristics of Bacteria	Forms spores; anaerobic	Does not form spores	Does not form spores; facultative; has survived freezing and high acidity (pHs below 4.0); can grow at refrigerator temperatures
Type of Illness	Intoxication (botulism)	Infection	Toxin-mediated infection
Symptoms	Vomiting and constipation or diarrhea may be present initially; progresses to fatigue, weakness, vertigo, blurred or double vision, difficulty speaking and swallowing, dry mouth; eventually leading to paralysis and death	Diarrhea (watery or bloody); fever, nausea, and vomiting; abdominal pain, headache, and muscle pain	Diarrhea (watery, may become bloody); severe abdominal cramps and pain; vomiting, and mild or no fever; may cause kidney failure in some people; symptoms more severe in the very young
Incubation Period	4 hours to 8 days; usually 12 to 36 hours	Usually 2 to 5 days, with a range of 1 to 10 days	3 to 8 days; usually 3 to 4 days
Duration of Illness	Several days to 1 year	2 to 5 days, usually no more than 10 days (relapses common); self-limiting	2 to 9 days
Source	Present on almost all foods of either animal or vegetable origin; soil; water	Domestic animals (sheep, pigs, cattle, poultry, and pets); the intestinal tracts of wild animals	Animals; particularly found in the intestinal tracts of cattle and humans
Foods Involved in Outbreaks	Foods that were underprocessed or temperature abused in storage; canned low-acid foods, untreated garlic-and-oil products, sautéed onions in butter sauce; leftover baked potatoes; stews; meat/poultry loaves; potential risks for MAP (modified-atmosphere packaging) and sous vide products	Unpasteurized milk and dairy products; raw poultry; nonchlorinated or fecal-contaminated water	Raw and undercooked ground beef; imported cheeses; unpasteurized milk and apple cider/juice; roast beef; dry salami; commercial mayonnaise; lettuce; nonchlorinated water
Preventive Measures	Do not use home-canned products; use careful time and temperature control for sous vide items and all large, bulky foods; purchase only acidified garlic-and-oil mixtures and keep refrigerated; sauté onions to order; rapidly cool leftovers	Thoroughly cook food to minimum safe internal temperatures; pasteurize milk; use treated water; avoid cross-contamination	Thoroughly cook ground beef to at least 155°F (68°C) for 15 seconds; avoid cross-contamination; practice good personal hygiene

Major Foodborne Illnesses Caused by Bacteria (Continued)

Foodborne Illness	Vibrio spp. (noncholerae) Gastroenteritis/Septicemia	Yersiniosis
Bacteria	Vibrio parahemolyticus and Vibrio vulnificus	Yersinia enterocolitica
Characteristics of Bacteria	Does not form spores; more common in warmer months	Does not form spores; facultative; can survive at pH below 4.5; can grow at low temperatures
Type of Illness	Infection	Infection
Symptoms	Diarrhea, abdominal cramps, nausea, vomiting, headache; severe cases of V. vulnificus include sudden chills, fever, blistering skin lesions, decreased blood pressure, and septicemia; fatality from V. vulnificus is high in immunocompromised people	Fever and severe abdominal pain (mimics appendicitis); possible diarrhea, headache, sore throat, or vomiting
Incubation Period	V. parahaemolyticus: 4 to 96 hours, usually 12 to 24 hours; V. vulnificus: 12 hours to several days, usually 38 hours	1 to 11 days; usually 24 to 48 hours
Duration of Illness	V. parahaemolyticus: 1 to 8 days; V. vulnificus: days to weeks; death within a few days for immunocompromised people	Days to weeks; chronic in some individuals
Source	Oysters and other shellfish, (shrimp, lobsters, clams, and mussels), especially from the Gulf of Mexico	Domestic pigs primary reservoir; soil; water; wild animals; rodents
Foods Involved in Outbreaks	Raw or partially cooked oysters; raw or undercooked shellfish (clams and mussels)	Meats (pork, beef, lamb); oysters; fish; raw milk; contaminated pasteurized milk; tofu; nonchlorinated water
Preventive Measures	Avoid serving raw or undercooked seafood (particularly oysters); avoid cross-contamination; freezing does not completely destroy these bacteria	Minimize cross-contamination from pork; thoroughly cook foods to minimum safe internal temperatures; ensure that facilities and equipment are properly sanitized; follow proper storage procedures; use only sanitary, chlorinated water supplies

Multiple Barriers for Controlling the Growth of Microorganisms

Multiple barriers need to be put in place, which will deny the microorganisms as many of the conditions that support growth as possible. The list below provides some simple barriers that can be combined to create multiple barriers.

○ **Make the food more acidic.** Add vinegar, lemon juice, lactic acid, or citric acid.

○ **Raise or lower the temperature of the food.** Move food out of the temperature danger zone by cooking it to the proper temperature, or refrigerating it to 41°F (5°C) or below, or by freezing it.

○ **Lower the water activity (a$_w$) of the food.** Dry food by adding sugar, salt, alcohol, or acid. Food can also be air dried or freeze dried to remove water.

○ **Lessen the amount of time the food is in the temperature danger zone.** Prepare food as close to service as possible.

VIRUSES

Viruses are the smallest of the microbial contaminants. While a virus cannot reproduce outside a living cell, once inside, it will reproduce more viruses. Viruses are responsible for several foodborne illnesses such as Hepatitis A, Norwalk Virus, and Rotavirus.

Basic Characteristics of Viruses

Viruses share some basic characteristics.

○ They rely on a living cell to reproduce.

○ They do not require a potentially hazardous food as a medium for transmission.

○ They are not complete cells.

○ They do not reproduce in food.

○ Some may survive freezing and cooking.

○ They can be transmitted from person to person, from people to food, and from people to food-contact surfaces.

○ They usually contaminate food through a foodhandler's improper personal hygiene.

○ They can contaminate both food and water supplies. An example is shellfish harvested from sewage-contaminated waters.

○ Practicing good personal hygiene is an important defense against foodborne viruses. It is especially important to minimize hand contact with ready-to-eat foods.

Major foodborne illnesses caused by viruses and methods of prevention are provided on the following page.

Major Foodborne Illnesses Caused by Viruses

Foodborne Illness	Hepatitis A	Norwalk Virus Gastroenteritis	Rotavirus Gastroenteritis
Virus	*Hepatovirus* or Hepatitis A	Norwalk and Norwalk-like viral agents	Rotavirus Gastroenteritis
Type of Illness	Infection	Infection	Infection
Symptoms	Mild or no illness, then sudden onset of fever, general discomfort, fatigue, headache, nausea, loss of appetite, vomiting, abdominal pain, and jaundice after several days	Nausea, vomiting, diarrhea, abdominal cramps, headache, and mild fever	Vomiting and diarrhea, abdominal pain, and mild fever (illness more common in children than adults)
Incubation Period	10 to 50 days	Usually 1 to 2 days; ranges from 10 to 50 hours	Usually 1 to 3 days
Duration of Illness	1 to 2 weeks; severe cases may last several months	Usually 1 to 3 days	Usually 4 to 8 days
Source	Human intestinal and urinary tracts; contaminated water	Human intestinal tract; contaminated water	Human intestinal tract; contaminated water
Foods Involved in Outbreaks	Water; ice; shellfish; salads; cold cuts and sandwiches; fruits and fruit juices; milk and milk products; vegetables; any food that will not receive a further heat treatment	Water; shellfish (especially raw or steamed); raw vegetables; fresh fruits and salads; contaminated water	Water and ice; raw and ready-to-eat foods (such as salads, fruits, and hors d'oeuvres); contaminated water
Preventive Measures	Obtain shellfish from approved sources; prevent cross-contamination from hands; ensure foodhandlers practice good personal hygiene; clean and sanitize food-contact surfaces; and use sanitary water sources	Obtain shellfish from approved sources; prevent cross-contamination from hands; ensure foodhandlers practice good personal hygiene; thoroughly cook foods to minimum safe internal temperatures; and use sanitary, chlorinated water	Ensure foodhandlers practice good personal hygiene; thoroughly cook foods to minimum safe internal temperatures; use sanitary, chlorinated water

PARASITES

Parasites share some basic characteristics.

○ They are living organisms that need a host to survive. A host is a person, animal, or plant on which another organism lives and takes nourishment.

○ They grow naturally in many animals such as hogs, cats and rodents and can be transmitted to humans.

○ Most are very small, often microscopic, but larger than bacteria.

○ They may be killed by proper cooking or freezing.

○ They pose hazards to both food and water.

Major foodborne illnesses caused by parasites and methods of prevention are presented on the next two pages.

FUNGI

Fungi range in size from microscopic, single-celled organisms to very large, multicellular organisms. They are found naturally in air, soil, plants, animals, water, and some foods. Molds, yeasts, and mushrooms are examples of fungi.

Molds

Molds share some basic characteristics.

○ They spoil food, and sometimes cause illness.

○ They grow under almost any condition, but grow well in sweet, acidic foods, with low water activity.

○ Freezing temperatures prevent or reduce the growth of mold but do not destroy it.

○ Some molds produce toxins called aflatoxins.

○ Mold cells and spores may be killed by heating, but mold toxins may not be destroyed.

○ To avoid mold toxins, throw out moldy food, unless the mold is a natural part of the food (e.g., cheeses such as Gorgonzola, bleu cheese, Brie, and Camembert. These are specific types of molds, which are grown in very specific conditions).

Although mold cells and spores can be killed by heating them, toxins that may be present are not destroyed by normal cooking methods. Foods with molds that are not a natural part of the product should always be discarded.

Major Foodborne Illnesses Caused by Parasites

Foodborne Illness	Trichinosis	Anisakiasis	Giardiasis
Parasite	*Trichinella spiralis*	*Anisakis simplex*	*Giardia duodenalis* (formerly *lamblia*)
Characteristic of Parasites	Roundworm	Roundworm	Protozoan
Type of Illness	Infection	Infection	Infection
Symptoms	Nausea, diarrhea, abdominal pain, occasionally vomiting, swelling around the eyes, fever, later, muscle soreness, thirst, extreme sweating, chills, hemorrhaging (bleeding), and fatigue	Tingling or tickling sensation in throat, vomiting or coughing up worms; in severe cases, severe abdominal pain, cramping, vomiting, and nausea	Fatigue, nausea, intestinal gas, weakness, weight loss, and abdominal cramps
Incubation Period	2 to 28 days; depends on number of larvae ingested	A few hours to 2 weeks	3 to 25 days; usually about 7 to 10 days
Duration of Illness	Several days to more than 30 days; depends on treatment and health status of individual	Up to 3 weeks	Usually 1 to 2 weeks; infectivity may last months
Source	Domestic pigs; wild game, such as bears and walruses	Marine fish, especially bottom-feeders	Wild animals; especially beavers and bears; domestic animals (dogs and cats); intestinal tract of humans
Foods Involved in Outbreaks	Undercooked pork or wild game; pork and nonpork sausages (ground meats may be contaminated by meat grinders)	Raw, undercooked, or improperly frozen seafood, especially cod, haddock, fluke, Pacific salmon, herring, flounder, monkfish, and fish used in sushi and sashimi	Water; ice; salads; (possibly) other raw vegetables
Preventive Measures	Cook pork and other meats to minimum internal cooking temperatures; wash, rinse, and sanitize equipment such as sausage grinders and utensils used in the preparation of raw pork and other meats	Obtain seafood only from certified sources; properly freeze fish; avoid eating raw or partly cooked fish and shellfish; fish intended to be eaten raw should be frozen at -4°F (-20°C) or lower for 7 days in a freezer, or at -31°F (-35°C) or lower for 15 hours in a blast chiller; *Anisakis* parasites are not killed by marinades	Use sanitary, chlorinated water supplies; ensure that foodhandlers practice good personal hygiene; wash raw produce carefully

Major Foodborne Illnesses Caused by Parasites (Continued)

Foodborne Illness	Toxoplasmosis	Intestinal Cryptosporidiosis	Cyclosporiasis
Parasite	*Toxoplasma gondii*	*Cryptosporidium parvum*	*Cyclospora cayetanensis*
Characteristic of Parasites	Protozoan; not directly passed from person to person	Protozoan; sporocysts are resistant to most chemical disinfectants	Protozoan
Type of Illness	Infection	Infection	Infection
Symptoms	Often, there are no symptoms. When symptoms occur they include enlarged lymph nodes in head and neck; severe headaches, severe muscle pain; rash. Individuals with compromised immune systems, such as HIV-infected people, and pregnant women and their fetuses, are most at risk.	Severe, watery diarrhea; may have no symptoms	Watery diarrhea, loss of appetite, weight loss, bloating, gas, abdominal cramps, nausea, vomiting; muscle aches; mild or no fever; fatigue
Incubation Period	5 to 20 days; depends on immune system of individual	1 to 12 days; average is 7 days	Days to weeks; usually about 1 week
Duration of Illness	Depends on treatment (relapses common)	4 days to 3 weeks; indefinite and life-threatening in immunocompromised people	A few days to a month, or longer; usually several weeks (relapses spanning 1 to 2 months are common)
Source	Animal feces (especially felines), mammals, birds	The intestinal tract of humans, cattle, and other domestic animals; drinking water contaminated with runoff from farms or slaughterhouses	The intestinal tract of humans; contaminated water supplies
Foods Involved in Outbreaks	Raw or undercooked meat contaminated with this parasite, especially pork, lamb, venison, and hamburger meat	Water; salads and raw vegetables; milk; raw foods, such as apple cider; ready-to-eat foods	Water; marine fish; raw milk; raw produce
Preventive Measures	Avoid raw or undercooked meats; thoroughly cook meats to the minimum internal temperature, so there is no pink inside; properly wash hands that come in contact with soil, raw meat, cat feces, or raw vegetables	Ensure that foodhandlers practice good personal hygiene; thoroughly wash produce; use sanitary water	Ensure that foodhandlers practice good personal hygiene; thoroughly wash produce; use sanitary water

Yeasts

Some yeasts are known for their ability to spoil food rapidly. Carbon dioxide and alcohol are produced as yeast slowly consumes food. Yeast spoilage may therefore produce an alcoholic smell or taste. Yeast may appear as a pink discoloration or slime and may bubble.

Yeasts are similar to molds in that they grow well in sweet, acidic foods with low water activity, such as jellies, jams, syrup, honey, and fruit juice. Foods that have been spoiled by yeast should be discarded.

FOODBORNE INFECTION VERSUS FOODBORNE INTOXICATION

Foodborne diseases are classified as infections, intoxications, or toxin-mediated infections.

○ **Foodborne Infections**
  - Result when pathogens grow in the intestines (after a person eats food contaminated by them).
  - Symptoms of a foodborne infection do not typically appear immediately.

○ **Foodborne Intoxications**
  - Result when food containing toxins is eaten. A person does not need to eat live organisms to become ill, just the toxins produced.
  - Symptoms of a foodborne intoxication typically appear quickly, within a few hours.

○ **Foodborne Toxin-Mediated Infections**
  - Result from eating food that contains pathogens. These pathogens grow in the intestines and produce toxins that can make you ill.

SUMMARY

Microbial contaminants are responsible for the majority of foodborne illness. Understanding how microorganisms grow, reproduce, contaminate food, and infect humans is critical to understanding how to prevent the foodborne illnesses they cause.

Of all foodborne microorganisms, bacteria are of greatest concern to the manager. They are more commonly involved in foodborne illness than all other foodborne microorganisms. Under favorable conditions, they can reproduce very rapidly. Although vegetative bacteria may be resistant to low and even freezing temperatures, they can be killed by high temperatures such as those reached during cooking. Some types of bacteria, however, have the ability to change into a spore, which protects the bacteria from unfavorable conditions. Since spores are so difficult to destroy, it is important to properly cook, cool, and reheat foods in order to keep bacteria from growing to harmful levels.

The acronym FAT-TOM is the key to controlling the growth of micro-organisms. Multiple barriers need to be put in place, which will deny the microorganism as many of the conditions that support growth as possible. These include making food more acidic, lowering its temperature, and lowering the water activity of the food.

Viruses are the smallest of the microbial contaminants. While a virus cannot reproduce in food, once ingested it will cause illness. Practicing good personal hygiene and minimizing hand contact with ready-to-eat foods is an important defense against foodborne illness from viruses.

Parasites are organisms that need to live in or on a host organism to survive. They can live inside many animals that humans eat, such as cows, chickens, hogs, and fish. They may be killed by proper cooking and freezing.

Molds and yeasts are examples of fungi, which are a concern to establishments. Molds are mostly responsible for the spoilage of many foods but can produce toxins, which can harm us. Foods with molds that are not a natural part of the product should always be discarded. Yeasts are known for their ability to spoil food rapidly. Foods that have been spoiled by yeast should be discarded.

Foodborne diseases are classified as infections, intoxications, or toxin-mediated infections. Foodborne infections result when pathogens grow in the intestines (after a person eats food contaminated by them). Foodborne intoxications are caused by eating food containing toxins. Foodborne toxin-mediated infections result from eating food that contains pathogens. These pathogens grow in the intestines and produce toxins that can make you ill.

ACTIVITY

Crossword Puzzle

Across:

2. Disease-causing microorganisms.

4. The amount of moisture in food available for the growth of microorganisms.

5. The temperature range between 41°F and 140°F (5°C to 60°C) within which most bacteria grow and reproduce.

6. This is caused by eating food containing poisonous toxins.

7. Those bacteria that are in the process of reproduction.

11. Mushrooms are an example of these.

12. A measure of a food's acidity or alkalinity.

14. This results when pathogens grow in the intestines (after a person eats food contaminated by them).

Down:

1. A type of fungus that causes food spoilage.

3. This results when pathogens grow in the intestines and produce toxins.

7. The smallest of the microbial food contaminants.

8. A type of fungus that causes food spoilage and sometimes cause illness.

9. An alternative form for some bacteria that protects them from adverse conditions.

10. Small, living organisms that can be seen only with the aid of a microscope.

12. An organism that needs to live in or on a host organism to survive.

13. Single-celled living microorganisms that can cause food spoilage and disease.

15. An acronym for the conditions needed by most microorganisms to grow.

ACTIVITY

Word Find

Find the terms that go with the clues below.

Clues

1. Small, living organisms that can be seen only with the aid of a microscope.

2. An organism that needs to live in or on a host organism to survive.

3. A measure of a food's acidity or alkalinity.

4. An alternative form for some bacteria that protects them from adverse conditions.

5. This results when pathogens grow in the intestines and produce toxins.

6. Those bacteria that are in the process of reproduction.

7. Single-celled living microorganisms that can cause food spoilage and disease.

8. The smallest of the microbial food contaminants.

9. This results when pathogens grow in the intestines (after a person eats food contaminated by them).

10. Disease-causing microorganisms.

11. Mushrooms are an example of these.

12. This is caused by eating food containing poisonous toxins.

13. An acronym for the conditions needed by most microorganisms to grow.

14. The amount of moisture in food available for the growth of microorganisms.

15. A type of fungus that causes food spoilage and sometimes causes illness.

16. A type of fungus that causes food spoilage.

17. The temperature range between 41°F and 140°F (5°C to 60°C) within which most bacteria grow and reproduce.

```
S P N O I T C E F N I E N R O B D O O F M V P M
F H S I H L V V T G O S R W Q D J X Y O I A S H
T A P H L J P X Z F Y C X Z X K K G L R T I A N
N O T E Z F R D M M E H U B D O N D U H N S Y O
A F X T R X Z K B K F E B I J G C S O A F E A I
T F D I O O S H K N I N S I J J S G G K J S D T
R D W O N M P R S F L B T N W Y E R B B G Z K A
O E P O V M O S K J J H D B R N O M F D Q B D C
P Q N C K E E K S E G M S W S O P F U S K Z T I
K L T L E Y O D J R Z B S X R L P I O A P C M X
Y K K P I V Q R I A I A Z C U K Q S A H D E S O
T W X J F H T I M A H T I P W S Y D U X Q Q I T
I Y P V N K K O Q I T M I D N G R K Y G P O N N
V K E N I C D J E P E E B X N J D J T L I E A I
I N O Z C P I D A V V A D H P H J W Q V X H G E
T T C B Y Q X R I P E G L I J N Z J L K L M R N
C C M O K F A T G K Q T J E N A N B W M I I O R
A B X Y U S A Z C Y G N U T W F I Z P G O H O O
R S I N I T M Y F P P A U G N U E R X K I R R B
E H G T E U U H A L K O G N E R E C E T N U C D
T I E G L G T P X L Z Q R L P J G J T T S B I O
A Z E Y W L E E X H Y P R J G Z A F A I C A M O
W V N U M D X U Y T V E Z R G U M H C U O A E F
E N O Z R E G N A D E R U T A R E P M E T N B Y
```

ACTIVITY

Who Am I?

Identify the illness from the characteristics given for each.

1. _____

- ○ I can be carried in the intestinal tract of humans.
- ○ I am sometimes found in sliced melons.
- ○ I can produce fever and diarrhea in those who ingest me.
- ○ My growth can be slowed by refrigeration.

2. _____

- ○ I most often affect pregnant women.
- ○ I am sometimes found in prepared and chilled ready-to-eat foods.
- ○ I can produce nausea and persistent fever in those who ingest me.
- ○ Proper cleaning and sanitizing can control me.

3. _____

- ○ I can be carried in the urinary tract of humans.
- ○ I am sometimes found in shellfish.
- ○ I can produce fatigue and a yellowing of the skin.
- ○ Obtaining shellfish from an approved source can be a safeguard against me.

4. _____

- ○ I can last three weeks.
- ○ I am sometimes found in sushi fish.
- ○ I can produce a tingling or tickling sensation in the throat.
- ○ Freezing sushi fish properly can destroy me.

5. _____

- ○ I can be carried in the intestinal tract of humans.
- ○ I am sometimes found in unpasteurized apple cider.
- ○ I can produce bloody diarrhea.
- ○ Thoroughly cooking ground beef can destroy me.

ACTIVITY

Word Jumble

Unscramble the letters to find the word that answers each clue. Then unscramble the highlighted letters in each word to solve the final clue.

1. TRACABIE: single-cell organism. (__)__ __ __ __ __ __ __

2. EAHSPOGL: stage where bacteria grows rapidly. (__)__ __   __ __ __ __ __

3. DOLM: spoilage microorganism. (__)__ __ __

4. DACICI: foods that have a pH between 0 and 7.0. __(__)__ __ __ __

5. PHOTANEGS: microbes that cause illness. __ __ __ __ __ __ __(__)__

6. TRAIPASE: a host's unwanted guest. __ __(__)__ __ __ __ __

7. NOTIFINEC: when microbes invade. __ __ __ __(__)__ __ __ __

8. GINFU: mold or yeast. __ __ __ __(__)

9. ROSPES: armored bacteria. __ __(__)__ __ __

10. URIVS: smallest of bugs. __(__)__ __ __

11. AMOTFT: key to microbial growth. __ __(__)–__ __ __

12. XINTOS: poisons. __ __ __ __(__)__

13. ESTAY: produces an alcoholic smell or taste. __ __(__)__ __

14. XGYENO: a condition of microbial growth. (__)__ __ __ __ __

15. OGRICSMOMSIARN: small beings. __ __ __ __ __ __ __ __ __ __ __ __(__)__

16. VITYCATWAIRET: food moisture level. __ __ __ __ __   (__)__ __ __ __ __ __ __

17. LTUERAN: this type of pH supports microbial growth. (__)__ __ __ __ __ __

18. HAPGLASE: bacterial adjustment period. __ __ __   __ __(__)__ __

19. IOATIXCNTNOI: eating a toxic food. (__)__ __ __ __ __ __ __ __ __ __

20. SOHT: parasite's home. __ __ __(__)

21. THESAHEAPD: the population of bacteria declines. __ __ __ __ __   __ __ __(__)__

Final Clue: What you don't want in your food. __ __ __ __ __ __ __ __ __

__ __ __ __ __ __ __ __ __ __ __ __.

ACTIVITY

FAT-TOM Strikes Again!

On Thursday, Carol's Catering delivered lunch for a meeting in a conference room at Acme Mfg. Carol and her partner put out a tropical fruit plate (with a pH of 2.5), a basket of breads and rolls (a_w of 7.0), and a deli tray with cheeses, sliced roast beef and roast pork, (with a pH of 5.0 and a_w of 0.98), and grilled chicken breasts (with a pH of 6.4 and a_w of 0.98) at 11:00 a.m. When lunch was over, meeting attendees put the leftover food on the counter in the pantry next to the meeting room. At 4:00 p.m., two employees made chicken sandwiches for themselves with the leftovers. Carol came back at 5:00 p.m. to clean up the conference room and collect the deli trays. The next day, the two employees called in sick and were later treated at the hospital for Salmonellosis.

Fill in the condition for each initial of the acronym FAT-TOM, and give reasons why these conditions supported the growth of Salmonella.

F _____. _____

A _____. _____

T _____. _____

T _____. _____

O _____. _____

M _____. _____

A CASE IN POINT

Case Study

A daycare center is serving stir-fried rice for lunch. The rice was properly cooked at 1:00 p.m. the day before; that is, it was heated to the minimum required temperature for at least fifteen seconds. The covered rice was then placed on the countertop and allowed to cool to room temperature. At 6:00 p.m., the cook placed it in the refrigerator. At 9:00 a.m. the following day, the rice was combined with the other ingredients for stir-fried rice and cooked to 165°F (74°C) for at least fifteen seconds. The cook covered the stir-fried rice and left it on the range until she gently reheated it at noon. Within an hour of eating the stir-fried rice, however, several of the children began vomiting, and a few had diarrhea. Samples from some of the children revealed that the rice, not the other ingredients, was considered to be the cause of the outbreak.

Based on the information given, was the illness caused by a bacteria, virus, parasite, or fungi? What is the name of the microorganism most likely to have caused the outbreak? Is this illness an infection or intoxication?

TRAINING TIPS

Training Tips for the Classroom

1. *Twenty Questions*

Objective: *After completing this activity, class participants should be able to identify the major foodborne illnesses caused by bacteria, viruses, and parasites, and be able to identify characteristics, symptoms, sources, foods involved, and preventive measures for each of these pathogens.*

Directions: After discussing Section 2, break your class into several groups. Assign each group a pathogenic bacteria, virus, or parasite that was discussed in the chapter.

Have each group study their "bug" for ten minutes. Ask them to focus on the following:

○ characteristics

○ symptoms of illness caused by the "bug"

○ source

○ foods involved

○ preventive measures

Then ask each group one at a time to stand in front of the class and solicit up to twenty questions from the other groups. Each group in the audience asks one question and makes a guess as to what bug that group represents. When a group correctly identifies the "bug," points are awarded based on the number of unasked questions remaining. For example, if four questions were asked before a correct guess was made, the group making the correct guess would be awarded sixteen points (20 - 4 = 16).

In the unlikely event that no team can figure out what bug a group represents, the group presenting receives twenty points. Play until every team has gone to the front. The team with the most points wins!

Note: Allow up to forty minutes for this game. If time is an issue, select the six or so most common "bugs," and assign larger groups.

2. Bug Scramble

Objective: *After completing this activity, class participants should be able to identify the major foodborne illnesses caused by bacteria, viruses, and parasites.*

Directions: Create a word scramble game for your class using the different types of bacteria, viruses, parasites, and fungi discussed in Section 2. Scramble the name of each bug and write it on a transparency (one bug per transparency).

Example: suerec sullicab *(Bacillus cereus)*

Then, break your class into four or five groups, and give each group a bell. Ask each group to ring the bell when they have unscrambled the bug projected on the screen. If they are correct, they get twenty points. (If you choose, you can assign a certain number of points for each bug, based on its perceived difficulty. Otherwise, to keep it simple, you can just give twenty points for each one.)

To enhance the learning process, you could require the group that unscrambles the word correctly to give two or three characteristics of the bug, and give bonus points for each correct answer.

As usual, the team with the most points wins!

MULTIPLE-CHOICE STUDY QUESTIONS

1. To what does the acronym FAT-TOM refer?

   A. The major types of microorganisms that can cause foodborne illness

   B. The conditions that support the growth of microorganisms

   C. Human health hazards associated with foods that are high in fat and sodium

   D. The federal agency responsible for monitoring food safety

2. All of the following are conditions that support the growth of microorganisms except

   A. moisture.

   B. a protein or carbohydrate food source.

   C. high acidity.

   D. temperatures between 41°F and 140°F (5°C and 60°C).

3. A person who has a foodborne infection has most likely eaten a food containing

   A. a ciguatoxin.   C. a plant toxin.

   B. histamine.   D. a live microorganism.

4. Which food is more likely to transmit parasites to humans?

   A. Improperly cooked eggs

   B. Improperly frozen sushi

   C. Improperly refrigerated milk

   D. Unpasteurized apple juice

5. Which of the following is not a basic characteristic of foodborne mold?

   A. It grows well in sweet acidic foods with low water activity.

   B. Freezing temperatures prevent or reduce its growth but do not destroy it.

   C. Its cells and spores may be killed by heating, but the toxins it produces may not be destroyed.

   D. It needs a host to survive.

6. Which of the following statements regarding foodborne intoxication is true?

A. The symptoms of intoxication often appear days after exposure.
B. The medical treatment for intoxication can be painful.
C. Foodborne intoxication is more common than foodborne infection.
D. The symptoms of intoxication appear quickly, within a few hours.

7. Which of the following microorganisms is most likely to cause a foodborne infection?

A. *Listeria monocytogenes*
B. *Bacillus cereus*
C. *E. coli* O157:H7
D. *Clostridium botulinum*

8. Which of the following microorganisms is most likely to cause a foodborne intoxication?

A. *Staphylococcus aureus*
B. *Shigella*
C. *Campylobacter jejuni*
D. *Salmonella*

9. Which of the following foods is commonly associated with an outbreak of *Bacillus cereus?*

A. Cooked rice
B. Blanched vegetables
C. Raw chicken
D. A hamburger cooked rare

10. Which of the following foods is commonly associated with an outbreak of *Vibrio?*

A. Raw oysters
B. Unpasteurized milk
C. Apple cider
D. Sliced melons

Section 3
Contamination, Food Allergies, and Foodborne Illness

Learning Essentials

After completing this section, you should be able to:

○ Identify the three types of contamination (biological, chemical, and physical), and give examples of each.

○ Identify ways in which food can become contaminated.

○ Identify methods to prevent biological, chemical, and physical contamination.

Knowledge

TEST YOUR FOOD-SAFETY KNOWLEDGE

1. **True or False:** Fish that has been properly cooked will be safe to eat. *(See Seafood Toxins, page 3-3.)*

2. **True or False:** Cooking can destroy fungal toxins in certain varieties of wild mushrooms. *(See Mushroom Toxins, page 3-4.)*

3. **True or False:** Food additives listed on the government's GRAS (Generally Regarded As Safe) list will not cause allergic reactions. *(See Monosodium Glutamate, page 3-7.)*

4. **True or False:** Unopened cleaning products may be stored with unopened cans and packages of food. *(See Chemicals and Pesticides, page 3-4.)*

5. **True or False:** Most biological toxins found in food occur naturally and are not caused by the presence of microorganisms. *(See Biological Contamination, page 3-3.)*

CONCEPTS:

○ **Biological contaminants:** A microbial contaminant that may cause foodborne illness. These contaminants may occur due to the introduction of bacteria, viruses, parasites, and fungi to food. Some biological toxins can also contaminate food.

○ **Ciguatera poisoning:** An illness that occurs when a person eats fish that have consumed the ciguatera toxin. This toxin occurs in certain predatory tropical reef fish such as amberjack, barracuda, grouper, and snapper.

○ **Scombroid poisoning:** An illness that occurs when a person eats a scombroid fish that has been time-temperature abused. Scombroid fish include tuna, mackerel, bluefish, skipjack, and bonito.

○ **Biological toxins:** A poison that is produced by a plant, or produced or ingested by an animal.

○ **Chemical contaminants:** A type of contaminant that causes foodborne illness. Food can become contaminated by a variety of chemical substances normally found in establishments including toxic metals, pesticides, and chemicals.

○ **Physical contaminants:** A type of contaminant that results from the accidental introduction of foreign objects into foods.

INTRODUCTION

Food is considered contaminated when it contains hazardous substances. These substances may be biological, chemical, or physical. The most common food contaminants are the biological contaminants that belong to the microworld—bacteria, parasites, viruses, and fungi, which were discussed in Section 2. Most foodborne illnesses are the result of microbial contamination, but biological and chemical toxins are responsible for many foodborne illnesses as well. Each year in the United States, it is estimated that biological and chemical toxins are responsible for millions of foodborne illnesses. While biological and chemical contamination pose a significant threat to food, the danger from physical hazards should also be recognized.

TYPES OF FOODBORNE CONTAMINATION

Ensuring the safety of food is the most important job of the manager. A thorough understanding of the causes and prevention of the various types of contamination can help you keep food safe.

Biological Contamination

As we learned in Section 2, a foodborne intoxication occurs when a person eats food containing a biological toxin which was produced or ingested by a plant or animal. Toxins in seafood, plants, and mushrooms are responsible for many cases of foodborne illness annually in the United States. Most of these toxins occur naturally and are not caused by the presence of microorganisms. Some occur in fish as a result of their diet.

Seafood Toxins

○ **Ciguatera toxins occur in certain predatory tropical reef fish such as amberjack, barracuda, grouper, and snapper.** Ciguatera accumulates in the tissue of these large, predatory fish after they eat smaller fish that have fed upon certain species of algae. Some important points about ciguatera include the following.

   ● Eating fish containing this toxin may result in an illness.

   ● Symptoms of ciguatera intoxication include vomiting, severe itching, nausea, dizziness, hot and cold flashes, temporary blindness, and sometimes hallucinations.

   ● Cooking does not destroy the ciguatera toxin, therefore it is very important to purchase predatory tropical reef fish only from approved suppliers.

○ **Shellfish may contain toxins that occur because of the algae upon which they feed.** The illnesses caused by shellfish poisoning vary and are specific to the type of toxin consumed.

   ● Since cooking may not destroy shellfish toxins, it is important to purchase shellfish from approved suppliers who can certify that they are harvested from safe waters.

○ **Scombroid poisoning is one of the more common forms of illness caused by fish toxins in the United States.** Some important points about scombroid poisoning include the following.

   ● It occurs when scombroid species of fish, such as tuna, mackerel, bluefish, skipjack, swordfish, and bonito are time-temperature abused. Under these conditions, the bacteria associated with the fish produce the toxin histamine.

   ● Symptoms of the illness include flushing and sweating, a burning peppery taste in the mouth, dizziness, nausea, and headache. Sometimes a facial rash, hives, edema, diarrhea, and abdominal cramps will follow.

   ● The histamine toxin is not destroyed by cooking or freezing.

   ● Since time-temperature abuse during the harvesting process may cause scombroid fish to become unsafe, it is important to purchase these fish from reputable suppliers who practice strict time-temperature controls.

To guard against seafood-specific foodborne illness, always purchase from reputable suppliers, and upon receipt verify that the product has been transported under strict temperature controls.

Mushroom Toxins

○ Outbreaks of foodborne illness associated with mushrooms are almost always caused by the consumption of wild mushrooms that have been collected by amateur mushroom hunters. Most cases occur when toxic mushroom species are confused with edible species.

○ The symptoms of intoxication vary depending upon the species consumed.

○ Cooking or freezing will not destroy the toxins produced by toxic mushrooms.

○ Establishments should not use wild mushrooms or products made with them. All mushrooms should be purchased from approved suppliers.

The table on page 3-5 summarizes information about common biological toxins.

Chemical Contamination

Chemical contaminants are responsible for many cases of foodborne illness. Contamination can come from a variety of substances normally found in establishments. These include toxic metals, pesticides, and chemicals.

Toxic Metals

○ Utensils and equipment that contain toxic metals such as lead, copper, brass, zinc, antimony, and cadmium can cause a foodborne illness from chemical contamination.

○ If acidic foods are stored in or prepared with this type of equipment, they can leach these metals from the item and contaminate the food.

○ Only food-grade utensils and equipment should be used to prepare and store food.

Chemicals and Pesticides

○ Chemicals such as cleaning products, polishes, lubricants, and sanitizers can contaminate food if they are improperly used or stored.

○ Follow the directions supplied by the manufacturer when using these chemicals.

○ To prevent contamination of food and food-preparation areas, use caution when using chemicals during operating hours.

○ Store chemicals away from food packaging, utensils, and equipment used for food. Keep them in a locked storage area in their original container.

Biological Contamination (Biological Toxins)			
Biological Toxins	**Source Of Contamination**	**Associated Foods**	**Preventive Measures**
Seafood Toxins			
Ciguatera Toxin	Fish that have eaten algae that contain the toxins	Predatory tropical reef fish such as amberjack, barracuda, grouper, and snapper	Cooking does not destroy these toxins; purchase tropical reef fish only from approved suppliers
Scombroid Toxin (Histamine)	Histamine produced by bacteria in some fish when they are time-temperature abused	Primarily occurs in tuna, bluefish, mackerel, skipjack, roundfish, and bonito; other fish, such as mahi-mahi, marlin, and sardines, have also been implicated in histamine poisoning	Cooking does not destroy histamine; because time-temperature abuse during the harvesting process may cause the fish to become unsafe, it is important to purchase from reputable suppliers
Shellfish Toxins	Shellfish that have eaten a type of algae that contains the toxin	Shellfish, especially mollusks, such as mussels, clams, cockles, and scallops	Cooking may not destroy this toxin; purchase these shellfish from approved suppliers who can certify that they are harvested from safe waters
Systemic Fish Toxins	Toxins that are a natural part of some fish	Pufferfish, moray eels, and freshwater minnows	Cooking may not destroy systemic fish toxins; pufferfish should be handled and prepared by properly trained chefs
Plant Toxins	Toxins that are a natural part of some plants	Fava beans, rhubarb leaves, jimsonweed, water hemlock, and apricot kernels; honey from bees that have gathered nectar from mountain laurel; milk from cows that have eaten snakeroot, jimsonweed, and other toxic plants	Cooking may not destroy these toxins; avoid these plant species and also products prepared with them
Fungal Toxins	Toxins that are a natural part of some varieties of fungi	Poisonous varieties of mushrooms and other fungi	Cooking does not destroy these toxins; do not use wild mushrooms; purchase mushrooms only from approved suppliers

○ If chemicals must be transferred to smaller containers or spray bottles, label each container appropriately.

○ Pesticides are often used in kitchens and in food-preparation and storage areas to control pests such as rodents and insects.

○ All food should be wrapped or stored prior to application of the pesticide. Pesticides should be stored with the same care as other chemicals used in the establishment.

The table on the page 3-6 summarizes information about some common chemical contaminants.

Chemical Contamination			
Chemical Toxin	**Source**	**Associated Foods**	**Preventive Measures**
Toxic Metals	Utensils and equipment containing potentially toxic metals such as lead, copper, brass, zinc, antimony, and cadmium	Any foods, but especially high-acid foods such as sauerkraut, tomatoes, and citrus products; the acidity of these foods can cause metal ions to leach into their liquids Carbonated beverages; the carbonated water used to make the beverage can leach copper ions from copper pipes into the water	Use only food-grade storage containers Use metal and plastic containers only for their intended use Use only foodservice brushes on food; do not use paintbrushes or wire brushes Do not use enamelware, which may chip and expose the underlying metal Do not cook high-acid foods in metal utensils Do not use equipment or utensils made of materials that contain lead (such as pewter) for food preparation Do not use zinc-coated (galvanized) equipment or utensils in food preparation Use a backflow prevention device to prevent carbonated water in soft drink dispensing systems from backflowing into the water intake system
Chemicals	Cleaning products, polishes, lubricants, and sanitizers	All foods	Follow manufacturers' directions for storage and use; use only recommended amounts Store away from food packaging, utensils, and equipment used for food Store in a dry cabinet in original labeled containers, apart from other chemicals that may react with them Tools used for dispensing chemicals should never be used on foods If chemicals must be transferred to smaller containers or spray bottles, label each container appropriately Use only food-grade lubricants or oils on kitchen equipment or utensils
Pesticides	Used in kitchens and food-preparation and storage areas to control pests, such as rodents and insects	All foods	Wash fruits and vegetables before preparation or consumption Pesticides should be applied only by a trained professional; wrap or store all food before applying pesticides

Physical Contamination

○ Physical contamination results from the accidental introduction of foreign objects into foods.

○ Common physical contaminants may include items such as metal shavings from cans, staples from cartons, glass from broken light bulbs, blades from plastic or rubber scrapers, fingernails, hair, bandages, and dirt.

○ Closely inspect the foods you receive and take steps to ensure that food will not be physically contaminated during the flow of food in your operation.

FOOD ALLERGIES

Some people are allergic to common food additives and preservatives such as sulfites, nitrites, and monosodium glutamate (MSG).

Physical Contamination
Metal shavings in an opened can may contaminate the food inside.

Nitrites

○ Nitrites are preservatives used by the meat industry.

○ Nitrites have been linked to cancer, especially when nitrite-treated meats are over-browned or burned, which produces cancer-causing substances.

Sulfites

○ Sulfites are used to preserve the freshness and/or color of certain foods such as dried and preserved fruits and vegetables.

○ Sulfites can cause allergic reactions which include nausea, diarrhea, asthma attacks, and in some cases, loss of consciousness.

Monosodium Glutamate (MSG)

○ Used to enhance flavor in many packaged foods; included on the federal government's GRAS (Generally Regarded as Safe) list of chemicals that are safe to use in food.

○ MSG can cause flushing, dizziness, headache, a dry burning throat, and nausea. Use only the recommended amount of MSG in recipes.

Your employees should be able to inform customers of menu items that contain these potential allergens. Designate one person per shift to answer customers' questions regarding your menu items. Keep the following points in mind to help customers with allergies enjoy a safe meal at your establishment.

○ Be able to fully describe each of your menu items when asked. Tell customers how the item is prepared and identify any "secret" ingredients that are used.

○ If you don't know if an item is free of an allergen, tell the customer. Urge the customer to order something else.

○ When preparing food for a customer with allergies, make sure that the food makes no contact with the ingredient that the customer is allergic to. Make sure all cookware, utensils, and tableware are allergen-free to prevent food contamination.

○ Serve menu items as simply as possible to customers with allergies. Sauces and garnishes are often the source of allergic reactions. Serve these items on the side.

SUMMARY

Biological and chemical toxins are responsible for many outbreaks of foodborne illness. Most occur naturally and are not caused by the presence of microorganisms. Some occur in the animal as a result of its diet. Since toxins are not living organisms, cooking or freezing typically will not destroy them. Most measures taken to prevent foodborne intoxications center on proper purchasing and receiving.

Purchase seafood from reputable suppliers who maintain strict time-temperature controls and can certify that the seafood has been harvested from safe waters. Do not use wild mushrooms or products made from them.

Use only food-grade utensils and equipment to prepare and store food. Cleaning products, polishes, lubricants, and sanitizers should be used as directed. Use caution when using these chemicals during operating hours and store them properly. If used, pesticides should be applied by a trained professional. Closely inspect the foods you receive and take steps to ensure that food will not become physically contaminated during its flow through your operation.

Some people are allergic to common food additives and preservatives such as sulfites, nitrites, and monosodium glutamate (MSG). You should be able to inform customers of these and other potential food allergens that may be included in the food that is served at your establishment.

ACTIVITY

Crossword Puzzle

Across:

3. A poison that is produced by a plant, or produced or ingested by a animal.

4. A poisoning that occurs when a person eats a scombroid fish that has been time-temperature abused.

5. Wrap foods before this substance is applied.

7. A microbial contaminant that may cause foodborne illness.

11. A pesticide is an example of this type of contaminant.

Down:

1. A type of contaminant that results from the accidental introduction of foreign objects into foods.

2. Flavor enhancer that sometimes causes allergies.

6. The toxin produced by bacteria in fish such as tuna and mackerel that causes scombroid intoxication.

8. A poisoning that occurs when a person eats fish that have consumed this toxin.

9. Over-browning or burning meats containing this preservative can create a cancer-causing substance.

10. Poisons.

ACTIVITY

Word Find

Find the terms that go with the clues below.

Clues

1. A microbial contaminant that may cause foodborne illness.

2. A poisoning that occurs when a person eats fish that have consumed this toxin.

3. A poisoning that occurs when a person eats a scombroid fish that has been time-temperature abused.

4. A poison that is produced by a plant, or produced or ingested by a animal.

5. A pesticide is an example of this type of contaminant.

6. The toxin produced by bacteria in fish such as tuna and mackerel that causes scombroid intoxication.

7. A type of contaminant that results from the accidental introduction of foreign objects into foods.

8. Over-browning or burning meats containing this preservative can create a cancer-causing substance.

9. Flavor enhancer that sometimes causes allergies.

10. Wrap foods before this substance is applied.

11. Poisons.

```
T P Z P S T N A N I M A T N O C L A C I S Y H P
F N E T O X I N S L B Z U O O C Y D A C H L N L
G B A S C Q R K O M Y J U F T J Z S A E D J Q U
Q I B N T W Y I P O Y N U G P V E L K K M Y K C
B O P U I I X Z T W L Z P L W X Z L W P U E I R
E L T X M M C G P P G I D D H O A K N P A G C D
T O J O X F A I H A L K J T E G U S J K U H S K
A G D V O E E T D S E L L T U V V D C A E I B H
M I H S Z A P R N E E P X A G E E N T M D B B K
A C V J E A G A X O S T E T V E M E I T M K P V
T A A S T R S V G B C B I M R U R C I Q E T G Q
U L A E U O S H E U C L H R L A A D E X G Z Q B
L T K C C A C N I F Z G A P T L I Y A E N A W R
G O T L U J U U G Z W X H C C I X L Y S Q P U J
M X F E J T C Q I B T M S O I H N N U A G Z P J
U I C M B C O L M G X C N R I G I J A G Y J M U
I N I G N M U Y L X O T J U Z Y O S P M M B R U
D Q A G R N W J A M A L J Z T B D L T F S V O Q
O A N O A A A X B M Z R J L M C R W O A U M V L
S F N E O J P R I N G R C B X Z O C K I M L A C
O O G U Q C O N S H Z E X C U G F C Q E B I S B
N U K F V I A X W G G A Z E L A G M R B C U N J
O N S L D N M S O A U K F U W L L U O Y H O E E
M I P F T Q C F T J B Y L U F A Y F V S F R D V
```

ACTIVITY

What's the Contaminant?

Write B, C, or P in the blank next to each contaminant to indicate whether it is biological, chemical, or physical.

1. Pesticides _____

2. Broken glass _____

3. Ciguatera _____

4. Lemonade stored in a galvanized container _____

5. Scombroid toxin _____

6. Metal shavings _____

7. Chlorine _____

8. Puffer fish _____

9. Copper _____

10. Fish bones _____

11. Salmonella _____

12. Wild mushrooms _____

13. Shellfish toxins _____

14. Hair _____

15. Plant toxins _____

ACTIVITY

Hazard Match

A. Match the food to the hazard it poses.

1. _____ Packaged food flavor enhancer   A. Physical contaminant

2. _____ Rhubarb leaves   B. Ciguatera poisoning

3. _____ Puffer fish   C. Allergy to sulfites

4. _____ Tomatoes cooked in copper pot   D. Scombroid poisoning

5. _____ Mahi-mahi   E. Plant toxin

6. _____ Shellfish   F. Allergy to MSG

7. _____ Processed meats   G. Systemic fish toxin

8. _____ Grouper   H. Shellfish poisoning

9. _____ Dried fruits   I. Allergy to nitrites

10. _____ Chicken bones   J. Toxic metals

B. For each food item, identify a measure to prevent the hazard from occurring.

1. _____

2. _____

3. _____

4. _____

5. _____

6. _____

7. _____

8. _____

9. _____

10. _____

ACTIVITY

You Draw the Conclusion

Scenario: In a small coastal town, a restaurateur ran into a few problems. Paul, the owner of Tropical Reef Hideaway, was faced with some customer complaints of illness. Paul and his staff are good at following food-safety practices. Paul's investigation into the complaints resulted in the following information. Your job is to assist Paul by asking questions to identify the possible cause of the illnesses.

○ The doors of the restaurant were left open several times during the day, which let a large number of flies into the establishment.

○ The drink of the day was a Mai Tai (a fresh fruit juice-based alcoholic beverage) which was made in large quantities.

○ Fresh fish is purchased daily.

○ The customers who became ill complained of nausea, dizziness, and severe itching.

○ All fish was cooked to the required minimum internal cooking temperature.

○ The customers ate raw clams on the half-shell as an appetizer.

A CASE IN POINT

Case Study

Roberto receives a shipment of frozen mahi-mahi steaks. The steaks are frozen solid at the time of delivery, and the packages are sealed and contain a lot of ice crystals. Roberto accepts the mahi-mahi steaks and thaws them at a refrigeration temperature of 38°F (3°C). The thawed fish steaks are then held at this temperature during the evening shift, and are cooked to order. The cooks follow all appropriate guidelines for preparing, cooking, holding, and serving the fish, monitoring time and temperature throughout the process. Unfortunately, these fish steaks are implicated in an outbreak of scombroid intoxication. Explain why this may have happened.

TRAINING TIPS

Training Tips for the Classroom

1. *Pick Your Poison*

Objective: *After completing this activity, class participants will be able to identify foods often associated with biological toxins, and determine preventive measures that can be taken to keep these toxins from causing foodborne illness.*

Directions: Using the biological contamination chart as a reference, create a restaurant menu that includes (but is not limited to) the list of foods often associated with biological toxins. Break the class into teams of two. Pass out a copy of the menu to each team and ask them to do the following:

1. Circle the menu items which are often associated with a biological toxin.

2. Determine preventive measures that can be taken to keep these toxins from causing a foodborne illness.

Ask the teams to present their ideas to the rest of the class. Discuss these ideas, and create a list of preventive measures for each menu item.

2. *"You Found What In Your Soup?"*

Objective: *After completing this activity, class participants will be able to identify common physical and chemical contaminants, and determine methods to prevent them from occurring.*

Directions: Have your students take out a blank sheet of paper. Ask them to list as many examples of physical contamination of food as they can think of in exactly two minutes.

When time has expired, randomly solicit examples of physical contamination. After each example is given, ask if anyone in the class has ever experienced this type of physical contamination before, as a foodhandler or customer. If so, ask the following questions:

○ How do you think it might have happened?

○ Could it have caused illness or injury?

○ If not, what other type of damage did it cause?

○ How would you have responded to such an incident?

○ How could it have been prevented from happening?

After the examples of physical contamination have been exhausted, repeat the process with examples of chemical contamination. Create a checklist of safe foodhandling practices and policies that will prevent both physical and chemical contamination from occurring in an establishment.

Training Tips on the Job

1. Developing a Customer Complaint Form

Purpose: *To document customer complaints regarding food items. Information on this form can then be used to identify corrective actions that may need to be made in the establishment's processes, procedures, or facility.*

Directions: Create a customer complaint form with the following headings:

Date   **Item**   **Reason Sent Back**   **Corrective Action**

Require employees to document incidents that resulted in the customer sending food back to the kitchen. Analyze customer complaints to determine if the complaint was the result of a hazard that requires a change in current processes or procedures.

Share customer complaints with employees, and get their input regarding corrective actions and preventive measures that should be taken. Explain to employees the importance of documenting customer complaints. Make sure that employees are not penalized for telling the truth, especially since the complaint may be the result of a mistake they made.

2. Performing a Chemical Hazard Inspection

Purpose: *To determine how food might become chemically contaminated in your facility and to identify corrective actions that must be taken.*

Directions: Using the sources of contamination identified in the chemical contamination chart, conduct an inspection of your facility to identify any possible problems. Solicit help from your foodservice team to do this. Look at the layout of the facility itself and the foodhandling procedures that are followed during receiving and storage, preparation and cooking, and holding and serving. Identify places where food can become physically or chemically contaminated. Look for possible worst-case scenarios.

Next, identify preventive measures that can be taken. Let the preventive measures identified in the chart serve as a guide. Some corrective actions can be made immediately (such as simple changes in procedures), while others may require more time (equipment changes).

Keep your foodservice team involved in this process, especially if changes will affect their work habits or current procedures. Everyone must be notified of any changes in processes or procedures, and the reasons for these changes must be clearly explained.

MULTIPLE-CHOICE STUDY QUESTIONS

1. You have ordered some frozen tuna steaks for your restaurant. When the delivery arrives, you notice that there is excessive frost and ice in the package. You refuse the delivery. Why?

   A. You suspect that the steaks may have been contaminated with a cleaning compound.

   B. You suspect the fish may contain the ciguatera toxin.

   C. You suspect that the steaks have been time-temperature-abused and may cause scombroid intoxication if you serve them.

   D. You believe the supply company may have treated the steaks with an unauthorized preservative.

2. Which food ingredient could cause stew to become chemically contaminated?

   A. Canned tomatoes   C. Dry mustard

   B. Cubed beef   D. Wild mushrooms

3. Your pest control company sprayed during the overnight closing hours. When you arrived the next morning, you noticed that some apples and grapes had been left out on the counter that night. What should you do?

   A. Discard the fruit.

   B. Wash the fruit under running water before serving it.

   C. Put the fruit in the refrigerator immediately.

   D. Call the pest control company to find out if they sprayed near the fruit.

4. You scoop some ice from your restaurant's ice storage bin and notice a drinking glass inside. The glass is missing a large piece of the rim. You find the chunk at the bottom of the storage bin. What should you do?

   A. Inspect each cube carefully by hand before serving the ice in beverages.

   B. Discard all of the ice and clean out the bin before restarting the icemaker.

   C. Transfer the ice to a clean container and flush out the bin before putting the ice back.

   D. No action is needed since you found both pieces of the glass.

5. Which situation could result in the physical contamination of a food-preparation area?

   A. A foodservice worker forgets to wear a hat while working in the area.
   B. The exhaust hood above the area is made of stainless steel.
   C. MSG, which was used to prepare a previous item, was spilled in the area.
   D. The cutting board used in the area is made of wood.

6. Which of the following fish is not associated with the ciguatera toxin?

   A. Barracuda   C. Snapper
   B. Grouper   D. Puffer fish

7. Which action should an establishment take to prevent the contamination of food?

   A. Do not use fresh fruits and vegetables that have been chemically treated while growing.
   B. Purchase food products from approved suppliers only.
   C. Keep high-acid foods separate from other types of foods.
   D. Store foods only in clean glass containers.

8. Which of the following items is an example of a physical food contaminant?

   A. A wild mushroom
   B. Dirt on a head of lettuce
   C. A time-temperature-abused tuna steak
   D. Sulfites used to preserve a package of dried apricots

9. All of the following can lead to the contamination of food except

   A. cooking tomato sauce in a copper pot.
   B. storing orange juice in a pewter pitcher.
   C. using a backflow prevention device on a carbonated beverage dispenser.
   D. serving fruit punch in a galvanized tub.

10. Which of the following statements is true about fish containing ciguatera or scombroid (histamine) toxins?

    A. Freezing will destroy these toxins.
    B. Cooking will not destroy these toxins.
    C. Proper receiving will prevent these toxins from making a person sick.
    D. Cooking will destroy these toxins.

Section 4
The Safe Foodhandler

Knowledge

TEST YOUR FOOD-SAFETY KNOWLEDGE

1. **True or False:** A foodhandler who doesn't look or act sick could still spread a foodborne illness. *(See How Foodhandlers Can Contaminate Food, page 4-3.)*

2. **True or False:** Wearing gloves is an acceptable substitute for handwashing. *(See Use of Gloves, page 4-6.)*

3. **True or False:** Foodhandlers must always wash their hands after smoking. *(See Handwashing, page 4-4.)*

4. **True or False:** A foodhandler who has been diagnosed with Salmonellosis cannot continue to work at an establishment while he has the disease. *(See Policies for Reporting Illness and Injury, page 4-9.)*

5. **True or False:** If a foodhandler receives a minor cut on her hand, she should be sent home immediately. *(See Policies for Reporting Illness and Injury, page 4-9.)*

Learning Essentials

After completing this section, you should be able to:

○ Explain the relationship between personal hygiene and food contamination.

○ Give examples of types of personal behaviors that contribute to food contamination.

○ List examples of clothing, accessories, and other items worn by employees that pose a threat to food safety.

○ Respond properly to foodhandler cuts, wounds, and sores to ensure food safety.

○ Identify procedures foodhandlers must follow when using gloves.

○ Identify diseases and foodborne pathogens that may be transmitted between humans through food.

○ Identify and respond to employee health problems that are a threat to food safety.

○ List and describe the policies that pertain to eating, drinking, and smoking while working with food.

CONCEPTS

○ **Gastrointestinal illness:** An illness relating to the stomach or intestine.

○ **Carrier:** A person who carries pathogens, yet never becomes ill himself.

○ **Infected lesion:** A wound or injury that is contaminated with a pathogen.

○ **Personal hygiene:** Sanitary health habits that include keeping the body, hair, and teeth clean; wearing clean clothes; and washing hands properly.

○ **Hand sanitizer:** A liquid used to lower the number of microorganisms on the surface of the skin. Hand sanitizers should be used after proper handwashing, not in place of it.

○ **Finger cot:** A protective covering used to cover a properly bandaged cut or wound on the finger.

○ **Hair restraint:** A device used to keep foodhandlers' hair away from food and to keep foodhandlers from touching their hair.

○ **Single-use gloves:** Disposable gloves used to provide a barrier between the hands and the food they come in contact with. Gloves should never be used in place of handwashing. Foodhandlers should wash their hands when putting on gloves and when changing to a fresh pair.

○ **Jaundice:** Yellowish color to the skin and eyes, which could indicate that a person has contracted Hepatitis A.

INTRODUCTION

Customers frequently judge an establishment by observing the appearance and behavior of the foodhandler serving them. Good personal hygiene is a critical protective measure against foodborne illness, and customers expect it. By establishing a personal hygiene program that includes specific policies, and by training and enforcing those policies, you can minimize your risk of causing foodborne illness and lost business.

It is ironic that people are both the cause and the victims of foodborne illness incidents. At every step in the flow of food through the operation, from receiving through final service, foodhandlers can contaminate food and cause customers to become ill. The manager who wants to provide safe food must build a sanitary barrier between the product and the people who prepare, serve, and consume it. This requires trained foodhandlers who possess the knowledge, skills, and attitude necessary to operate a safe food system.

HOW FOODHANDLERS CAN CONTAMINATE FOOD

Foodhandlers can contaminate food when:

○ They have been diagnosed with a foodborne illness.

○ They show symptoms of gastrointestinal illness (an illness relating to the stomach or intestine).

○ They have infected lesions (wounds or injuries).

○ They live with or are exposed to a person who is ill.

○ They touch anything that may contaminate their hands.

Even an apparently healthy person may be hosting dangerous pathogens. With some diseases, such as Hepatitis A, an extremely hearty virus, a person is at the most infectious stage of the disease for several weeks before symptoms appear. With other diseases, such as Salmonellosis, the *Salmonella* bacteria may remain in a person's system for months after all signs of infection have ceased. Some people are called carriers because they may carry pathogens, yet never become ill themselves.

Simple acts or personal behaviors can contaminate food. These include things like:

○ Nose picking

○ Rubbing an ear

○ Scratching the scalp

○ Touching a pimple or an open sore

○ Running fingers through the hair

○ Coughing and sneezing into the hand

○ Spitting in the establishment

Because of these factors, foodhandlers must pay close attention to what they do with their hands, and wash them often.

DISEASES NOT TRANSMITTED THROUGH FOOD

In recent years, the public has expressed growing concern over communicable diseases that are spread through intimate contact or by direct exchange of bodily fluids. Diseases such as AIDS (Acquired Immune Deficiency Syndrome) and Hepatitis B are not spread through food. These diseases are spread through blood transfusions, shared needles, and sexual contact.

COMPONENTS OF A GOOD PERSONAL HYGIENE PROGRAM

Good personal hygiene is key to the prevention of foodborne illness. Good personal hygiene includes:

○ Hygienic hand practices

○ Maintaining personal cleanliness

○ Wearing clean and appropriate uniforms and following dress codes

○ Avoiding unsanitary habits and actions

○ Maintaining good health

○ Reporting illnesses

Hygienic Hand Practices

Handwashing

While handwashing may appear fundamental, many foodhandlers fail to wash their hands properly and as often as needed. As a manager, it is your responsibility to train your foodhandlers and then monitor them to make sure that they are washing their hands properly and when necessary. Never take this simple action for granted.

To ensure proper handwashing in your establishment, train your foodhandlers to use the steps illustrated on the next page.

Approved hand sanitizers (liquids used to lower the number of microorganisms on the surface of the skin) or hand dips may be used after washing, but should never be used in place of proper handwashing. If hand sanitizers are used, foodhandlers should never touch food or food-preparation equipment until the hand sanitizer has dried.

Foodhandlers must wash their hands after the following activities.

○ After using the restroom

○ Before and after handling raw foods

○ After touching the hair, face, or body

○ After sneezing, coughing, or using a handkerchief or tissue

○ After smoking, eating, drinking, or chewing gum or tobacco

○ After using any cleaning, polishing, or sanitizing chemical

○ After taking out garbage or trash

○ After clearing tables or busing dirty dishes

○ After touching clothing or aprons

○ After touching anything else that may contaminate hands, such as unsanitized equipment, work surfaces, or wash cloths

1. Wet your hands with hot running water.

2. Apply soap.

3. Rub hands together for at least twenty seconds.

4. Clean under fingernails and between fingers.

5. Rinse hands thoroughly under running water.

6. Dry hands.

Proper Handwashing Procedure

Hand Maintenance

In addition to proper washing, hands need other regular care to ensure that they will not transfer contaminants to food. Remember the following when considering hand maintenance.

○ Food, filth, and harmful substances can get caught under both long and short fingernails.

○ Fingernails should be kept short and clean.

○ Nail polish, false fingernails, and acrylic nails may be difficult to keep clean and can break off into food; therefore, these should not be worn while handling food.

○ Cuts and sores on hands, including hangnails, should be treated and kept covered with clean bandages.

○ If hands are bandaged, clean gloves or finger cots (protective coverings) should be worn at all times to protect the bandage and to prevent it from falling off into food.

○ You may need to move the foodhandler to another job, where he or she will not handle food or touch food-contact surfaces, until the injury heals.

Use of Gloves

If used, gloves must never be used in place of handwashing. Foodhandlers must wash their hands before putting on gloves and when changing to a fresh pair. Gloves used to handle food are for single use only and should never be washed and re-used. Foodhandlers should change their gloves when necessary. Gloves should be changed:

○ As soon as they become soiled or torn.

○ Before beginning a different task.

○ At least every four hours during continual use, and more often when necessary.

○ After handling raw meat and before handling cooked or ready-to-eat foods.

Other Good Personal-Hygiene Practices

In addition to following proper hand-hygiene practices, your foodhandlers must maintain personal cleanliness. Foodhandlers should bathe or shower before work. They must also keep their hair clean. Oily, dirty hair can harbor pathogenic microorganisms, and dandruff may fall into food or onto food-contact surfaces.

Proper Work Attire

A foodhandler's attire plays an important role in the prevention of foodborne illness, so foodhandlers must observe strict dress-code standards. Dirty clothes may harbor disease-causing microorganisms and give customers a bad impression of your establishment.

Managers should make sure that foodhandlers observe the following guidelines regarding their attire.

○ **Wear a clean hat or other hair restraint.** A hair restraint will keep hair away from food and keep foodhandlers from touching their hair. Foodhandlers with long beards should also wear beard restraints.

○ **Wear clean clothing daily.** If possible, foodhandlers should put on their work clothes at the establishment.

○ **Remove aprons when leaving food-prep areas.** For example, aprons should be removed prior to taking out garbage or using the restroom.

○ **Wear appropriate shoes.**

○ **Remove jewelry prior to preparing or serving food or while working around food-preparation areas.** Jewelry can harbor microorganisms, can tempt foodhandlers to touch it, and may pose a safety hazard around equipment. Remove rings (except for a plain wedding band), bracelets, watches, earrings, necklaces, and facial jewelry (such as nose rings, etc.)

Check with your local regulatory agency regarding requirements. These requirements should be written policies, which should be consistently monitored and enforced. All potential employees should be presented with these policies prior to employment.

Policies Regarding Eating, Drinking, Chewing Gum, and Tobacco

Small droplets of saliva can contain thousands of disease-causing micro-organisms. In the process of eating, drinking, chewing gum and tobacco, and smoking, this saliva can be transferred to the foodhandler's hands or directly to the food that they are handling. For this reason, foodhandlers must follow strict policies regarding these activities.

Managers must implement the following policies in their establishments.

○ Foodhandlers must not smoke, or chew gum or tobacco, while preparing or serving food, while in food preparation areas, and while in areas used for equipment and utensil washing.

○ Foodhandlers must not eat or drink while in food-preparation areas or in areas used to clean utensils and equipment (with the exception of chefs properly tasting foods). Some jurisdictions allow employees to drink from a covered container with a straw. Check with your local regulatory agency.

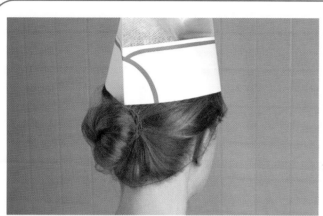

Hair properly restrained

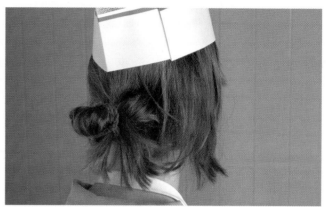

Hair improperly restrained

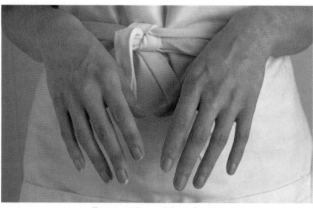

Proper hand hygiene:
clean, short fingernails; no jewelry or nail polish

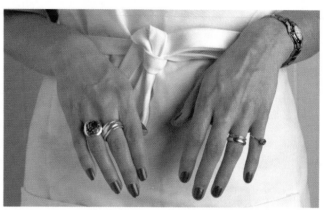

Improper hand hygiene:
long fingernails, jewelry, nail polish

Proper apron: clean

Improper apron: dirty and stained

Proper and Improper Attire on the Job

Policies for Reporting Illness and Injury

Foodhandlers must report health problems to the manager of the restaurant or establishment before working with food. If they become ill or are injured while working, they must report their condition to their manager or supervisor immediately. If the foodhandler's condition could possibly contaminate food or foodservice equipment, he or she must stop working and see a doctor. If the foodhandler must take medication while working, the medicine must be stored with his or her personal belongings away from areas where food is prepared, served, and stored.

According to the FDA Model Food Code, managers must exclude from the establishment foodhandlers diagnosed with a foodborne illness, and notify the local regulatory agency. Managers must also exclude foodhandlers from working with or around food if they have the following symptoms.

○ Fever

○ Diarrhea

○ Vomiting

○ Sore throat

○ Jaundice (yellow skin and eyes)

Managers must work with local regulatory agencies to determine when foodhandlers can safely return to work.

Any cuts, burns, boils, sores, skin infections, or infected wounds should be covered with a bandage when the foodhandler is working with or around food or food-contact surfaces. Bandages should be clean, dry, and must prevent leakage from the wound. As mentioned previously, waterproof disposable gloves or finger cots should be worn over bandages on hands. Foodhandlers wearing bandages may need to be temporarily reassigned to duties that do not involve contact with food or food-contact surfaces.

SUMMARY

Foodhandlers can contaminate food at every step in its flow through the operation. Good personal hygiene is a critical protective measure against contamination and foodborne illness. A successful personal hygiene program depends on trained foodhandlers who possess the knowledge, skills, and attitude necessary to maintain a safe food system.

Foodhandlers have the potential to contaminate food when they have been diagnosed with a foodborne illness, when they show symptoms of a gastrointestinal illness, when they have infected lesions, or when they touch anything that may contaminate their hands. Simple acts such as nose picking or running fingers through the hair can contaminate food. Proper handwashing must be practiced. This is especially important after using the restroom, before and after handling raw food, after sneezing and coughing, and after smoking, eating, or drinking. The manager needs to monitor handwashing to make sure it is thorough and frequent. In addition, hands need other care to ensure that they will not transfer contaminants to food. Fingernails should be kept short and clean. Cuts and sores should be covered with clean bandages, and covered with gloves or finger cots.

Foodhandlers must wash their hands before putting on gloves and when changing to a fresh pair. Gloves used to handle food are for a single use and should never be washed or re-used. They must be changed when necessary.

All employees must maintain personal cleanliness. They should bathe or shower before work and keep their hair clean. Prior to handling food, food-handlers must wear clean hair restraints and clothing, remove jewelry, and wear appropriate shoes. Aprons should always be removed when the employee leaves food-preparation areas.

Eating, drinking, smoking, and chewing gum and tobacco should not be allowed when the foodhandler is preparing or serving food or working in food-preparation areas. Foodhandlers must be encouraged to report health problems to management before working with food. If their condition could contaminate food or equipment, they must stop working and see a doctor. Managers must not allow foodhandlers diagnosed with a foodborne illness to work, and must notify the local regulatory agency. Managers must also exclude foodhandlers from working with or around food if they have symptoms that include fever, diarrhea, vomiting, sore throat, and jaundice.

ACTIVITY

Crossword Puzzle

Across:

2. A person who carries pathogens that may infect others with certain diseases, yet never becomes ill himself.

5. A disease not transmitted through food.

8. An illness relating to the stomach or intestine.

11. Yellowish color to the skin and eyes, which could indicate that a person has contracted Hepatitis A.

12. A protective covering used to cover a properly bandaged cut or wound on the finger.

13. An infected wound or injury.

14. Sanitary health habits that include keeping the body, hair, and teeth clean, wearing clean clothes, and washing hands properly.

Down:

1. A device used to keep foodhandlers' hair away from food and to keep the foodhandlers from touching their hair.

3. The key hygienic hand practice.

4. Disposable gloves used to provide a barrier between the hands and the food they come in contact with.

6. A liquid used to reduce the number of microorganisms on the surface of the skin.

7. Items that harbor microorganisms and may pose a safety hazard around equipment.

9. An item that should be removed when leaving food-preparation areas.

10. A personal behavior that can contaminate food.

ACTIVITY

Word Find

Find the terms that go with the clues below.

Clues

1. Sanitary health habits that include keeping the body, hair, and teeth clean; wearing clean clothes; and washing hands properly.

2. An illness relating to the stomach or intestine.

3. Yellowish color to the skin and eyes, which could indicate that a person has contracted Hepatitis A.

4. A person who carries pathogens that may infect others with certain diseases, yet never becomes ill himself.

```
I Z T I C I U G U S E V O L G E S U E L G N I S
N G S H A N D W A S H I N G Q N K A K G S F C Y
F H A D L A U S K O S D E V A H G W I B U R T A
E N N S I A B R Y G I C J C S E A E M W D S P F
C O O V T A L Y Z B O W L A I J U L N A W R I Q
T M S L D R E D Y Z Z R C O Y E N Y N Z O N C Y
E H E U G J O C X Q X W P V U I J O F N G R G D
D X P D M D P I I S K E M E P S Y L B E L B M Z
L W I J S G X Z N D Z X N F B R E A R E J D U E
E D C Y T D Q H N T N I P C T B W C N Y Q H V F
S S K F M H C E X E E U I P N E O E U D Y D O S
I R I D L J Q V M Z Y S A V F T I R M F S W I F
O E N C J L B S J J G Q T J K G Y B Z G N E A I
N Z G D B G G F X Q B M Y I Y S R W D O Q J B S
T I P E Z K V I J C V J C H N J S N C V Z D Z Z
E T A E H P Z K Z K D A L U F A E V M Q Z L N B
W I F A N Y Z R P B R A S Z H N L W T F P Y S Z
B N Q U O W F D W R N E Q C A B G I E S Y X Q J
Y A A A T X E W I O J S G H W V V R L L G R R A
Q S Y N Y I Q E S J Z T Y C Q F X K L L R B K E
X D Q V X Z R R H C B G P A O K V V R Y N Y K J
S N H A W V E F K X B N B H N K Y C H P C E I F
C A V C W P B Z D A X S D R X T P E R Y N S S C
V H P G Y P C U M P T N I A R T S E R R I A H S
```

5. Disposable gloves used to provide a barrier between the hands and the food they come in contact with.

6. A liquid used to reduce the number of microorganisms on the surface of the skin.

7. Items which harbor microorganisms and may pose a safety hazard around equipment.

8. A item that should be removed when leaving food-preparation areas.

9. A personal behavior that can contaminate food.

10. An infected wound or injury.

11. A protective covering used to cover a properly bandaged cut or wound on the finger.

12. A disease not transmitted through food.

13. A device used to keep the foodhandlers' hair away from food and to keep foodhandlers from touching their hair.

14. The key hygienic hand practice.

ACTIVITY

What's Wrong with this Picture?

A. The picture below shows a situation involving unsafe foodhandling. How many things can you find wrong with this picture?

B. Considering the unsafe practices noted above, develop guidelines that should be followed in this establishment.

ACTIVITY

Put the Steps in Order

Proper handwashing is essential to food safety. Foodborne illnesses are easily spread when employees don't wash their hands before handling food. Put the following steps in order to reveal proper handwashing technique.

_____   Rub hands together for at least twenty seconds.

_____   Wet your hands with hot running water.

_____   Rinse hands thoroughly under running water.

_____   Clean under fingernails and between fingers.

_____   Apply soap.

_____   Dry hands.

_____   Turn faucet off with a sanitary single-use paper towel.

ACTIVITY

Randall's Day

Randall is a foodhandler at a deli. It's 7:47 a.m. and Randall has just woken up. He is scheduled to be at work and ready to go by 8:00 a.m. When he gets out of bed his stomach feels queasy, but he blames that on the drinks he consumed the night before. Fortunately, he lives only five minutes from work, but he doesn't have enough time to take a shower. He grabs the same uniform that he wore the day before when he prepped chicken. Randall is wearing several pieces of jewelry from his night out on the town.

Randall doesn't have luck on his side today. En route to the restaurant his oil light comes on, and he is forced to pull off the road and add oil to his car. When he walks through the door at work he realizes that he has left his uniform hat at home. Randall is greeted by an angry manager, who puts him to work right away loading the rotisserie with raw chicken. Then he moves to serving a customer who orders a freshly made salad. Randall is known for his salads and makes the salad to the customer's approval.

The deli manager, who was short staffed on this day, asks Randall to take out the garbage and then prepare potato salad for the lunch hour rush. On the way back in, Randall mentions to the manager that his stomach is bothering him. But the manager, thinking of his short staff problems, asks him to stick it out as long as he can. Randall agrees and heads to the restroom in hope of relieving his symptoms. After quickly washing his hands in the restroom, he finds that the paper towels have run out. Short of time, he wipes his hands on his apron.

Later, while preparing the potato salad, Randall cuts his finger. He bandages the cut and continues his prep work. The manager summons Randall to clean the few tables that the deli has made available for customers. Randall puts on a pair of single-use gloves and cleans and sanitizes the tables. When finished, he grabs a piece of chicken from the rotisserie for a snack and immediately goes back to preparing the potato salad because it is almost noon.

A. Randall and the manager committed approximately nineteen errors in the above scenario. How many can you identify?

Targets:

○ If you can identify only eight to twelve errors, you may need to reread this section.

○ If you can identify thirteen to seventeen errors, you have a good understanding of this section.

○ If you can identify seventeen or more errors, you are on your way to becoming a health inspector.

B. Now that you have identified these errors, identify the policies that the deli manager should have had in place to prevent these errors from occurring.

ACTIVITY

Right Way/Wrong Way

Each pair of statements represents a right way and a wrong way to accomplish a task. Circle the right way to do each.

1A. When interrupted during food prep, remove your gloves, then put them back on when you return.
1B. When interrupted during food prep, remove your gloves; wash your hands and put on a fresh pair of gloves when you return.

2A. Remove your apron before using the restroom.
2B. Keep your apron on while using the restroom in case there are no paper towels.

3A. Put your uniform on at home to save time checking in before shift at work.
3B. Put uniform on at work if possible to prevent bringing contaminants into the establishment.

4A. Foodhandlers should cover cuts or sores on their hands with both a clean bandage and a finger cot or gloves before working around food.
4B. Foodhandlers can work around food when they have a cut or sore on their hands as long as it is covered with a clean, dry bandage.

5A. Foodhandlers who have to sneeze a lot because of a cold while working with food should step away from the food, use a tissue, wash hands, and change gloves before continuing.
5B. Foodhandlers who have to sneeze a lot because of a cold while working with food should stop working and go see a doctor.

6A. After using a restroom, foodhandlers can return to work after washing their hands properly for at least twenty seconds.
6B. After using restrooms, foodhandlers can return to work after proper handwashing, then washing their hands a second time when they return to their station.

7A. After preparing raw food with their bare hands, foodhandlers can serve food by putting on a pair of single-use gloves.
7B. After preparing raw food with their bare hands, foodhandlers must first wash their hands thoroughly before putting on a pair of single-use gloves to serve food.

A CASE IN POINT

Case Study

Marty works for a catering company. A few days ago, he was serving hot foods from chafing dishes at an outdoor music festival sponsored by the local community college. He did not wear gloves, because he used spoons and tongs to serve the food. His manager noticed that Marty made multiple trips to the bathroom during his four-hour shift. These trips did not interrupt service to customers because there were plenty of staff members and Marty hurried to and from the restroom.

The nearest restrooms had soap, separate hot and cold water faucets, and a working hot-air dryer, but no paper towels. Each time Marty used the restroom, he washed his hands quickly and then dried them on his apron. Throughout the following week, the manager of the catering company received several telephone calls from people who had attended the music festival and had eaten from the buffet. They each complained of diarrhea, fever, and chills. One call was from a mother of a young boy who was hospitalized for dehydration from the diarrhea. The doctor reported that the boy had Shigellosis.

Explain how Marty might have caused an outbreak of Shigellosis. What measures should have been taken to prevent such a foodborne-illness outbreak?

NOTES

TRAINING TIPS

Training Tips for the Classroom

1. Habits with Hands Role-Play

Objective: *After completing this activity, class participants should be able to determine how hands can cross-contaminate food, and identify the proper times to wash hands.*

Directions: Solicit three volunteers prior to teaching Section 4, and provide them with enough time to prepare for their role-play. Assign one volunteer to role-play a cook, another a server, and the third, a manager.

Volunteers are to design a three-minute role-play in which they portray their character performing various tasks. While doing so, they should also depict different ways to cross-contaminate food with their hands.

For example, the person playing the cook might handle raw poultry, and then prepare sandwiches without washing her hands, or the server may sneeze into his hands, wipe them on his apron, and then use his bare hands to serve dinner rolls to a guest. The goal is to design scenes with as many violations as possible.

Prior to presenting Section 4 to the class, tell them that a role-play is going to be performed. Ask class participants to note every incident of cross-contamination that they see during the role-play. At the end of the role-play, discuss the incidents noted, and develop a list that identifies when hands should be washed. Then discuss the proper procedure for handwashing.

2. Yuck

Objective: *After completing this activity, class participants should recognize the importance of proper handwashing.*

Directions: Purchase some petri dishes and petri-agar mix from a science supply store. Follow the directions for mixing the agar for each dish. Issue one petri dish per class participant. Ask each participant to lightly press a finger or thumb into the agar. Cover the petri dishes and place them in a warm location. Allow the dishes to sit for at least forty-eight hours. As a variation of this activity, you might ask participants to wash their hands and make another impression in the agar. Separate these impressions from the first set.

After a couple of days, a microbial growth should appear in the agar where the class participants left their finger or thumb impressions. Compare the impressions left by the unwashed hands to the impressions left by the washed hands.

This simple activity will show class participants that microorganisms, while invisible to the unaided eye, can be present on the body. It should also emphasize the importance of proper handwashing.

3. Glo Germ™ Activity

Objective: *After completing this activity, class participants should recognize the importance of proper handwashing.*

Directions: Ask for a volunteer from the class to participate in an experiment. Apply some iridescent "germs" (harmless orange liquid) from a *Glo Germ* kit to the hands of the volunteer and ask him to rub it in. Have him wash his hands.

When the volunteer returns, turn off the lights, and hold a black light over his hands to see how effectively he washed them. Often, the hands will glow around the fingernails and around watches or jewelry. Walk the volunteer around the room and use the black light to show the rest of the class how many germs have survived the handwashing process.

Discuss the importance of proper handwashing with the class.

Training Tips on the Job

1. Gotcha

Purpose: *This activity may help make handwashing a habit at your establishment.*

Directions: Create a list of activities after which handwashing must be performed. Share this list with all of your foodhandlers, and designate a "gotcha" day. The "gotcha" day will be a handwashing monitoring day. Foodhandlers will monitor one another. Whenever a foodhandler is caught not washing his or her hands after one of the listed activities, fellow foodhandlers should shout "gotcha."

By the end of the day it will be evident to the staff that handwashing is easy to forget when they are busy. Handwashing must become a habit at your establishment. Ask the staff to continue the "gotcha" activity until handwashing becomes a habit. If necessary, offer incentives to keep the staff involved in this activity.

2. Creating the Safe Foodhandler

Purpose: *To make foodhandlers aware of personal hygiene standards necessary to keep food safe in your establishment.*

Directions: Develop a colorful wall poster with a drawing of "The Safe Foodhandler." Make the drawing life-size if possible. Noted on the drawing should be all of the aspects of good hygiene from head to toe, from clean and properly restrained hair to proper closed-toe shoes. Make a checklist of basic hygiene standards such as coming to work healthy, taking a shower or bath before work, etc. Attach the checklist to the poster.

Place your poster in a conspicuous place for all foodhandlers to see.

MULTIPLE-CHOICE STUDY QUESTIONS

1. Which of the following personal behaviors can contaminate food?

   A. Touching a pimple   C. Nose picking

   B. Touching hair   D. All of the above

2. After you've washed your hands, which of the following items can you use to dry your hands?

   A. Your apron   C. A common cloth towel

   B. Single-use paper towels   D. A wiping cloth

3. Which item of personal apparel would most likely cause food to become unsafe?

   A. Earrings   C. A baseball-type cap

   B. A dark-colored shirt   D. A pair of athletic shoes

4. Which of the following is the proper procedure for washing your hands?

   A. Run hot water, moisten hands and apply soap, rub hands together, apply sanitizer, dry hands.

   B. Run hot water, moisten hands and apply soap, rub hands together, rinse hands, dry hands.

   C. Run cold water, moisten hands and apply soap, rub hands together, rinse hands, dry hands.

   D. Run cold water, moisten hands and apply soap, rub hands together, apply sanitizer, dry hands.

5. Kim wore disposable gloves while she formed raw ground beef into patties. When she was finished, she continued to wear the gloves while she sliced hamburger buns. What mistake did Kim make?

   A. She failed to change her gloves and wash her hands after handling raw meat and before handling a ready-to-eat food item.

   B. She failed to wash her hands before wearing the same gloves to slice the buns.

   C. She failed to wash and sanitize her gloves before handling the buns.

   D. She failed to wear reusable gloves.

6. Becky has an unhealed sore on the back of one hand. Can Becky perform her regular foodhandling duties?

A. Yes, if Becky's doctor provides a certificate that the sore is not contagious.

B. Yes, if Becky agrees to use an antiseptic hand lotion between jobs.

C. Yes, if the sore is bandaged and Becky wears a glove to protect the bandage.

D. No, Becky should stay home from work until the sore has healed.

7. Foodhandlers should be excluded from working with or around food if they are experiencing which of the following symptoms?

A. Fever, itching, fatigue   C. Vomiting, diarrhea, itching

B. Fever, vomiting, diarrhea   D. Fatigue, vomiting, itching

8. Which of the following policies should be implemented at establishments?

A. Employees must not smoke while preparing or serving food.

B. Employees must not eat while in food-preparation areas.

C. Employees must not chew gum or tobacco while preparing or serving food.

D. All of the above

9. Stephanie has a small cut on her finger and is about to prepare chicken salad. How should Stephanie's manager respond to the situation?

A. Send Stephanie home immediately.

B. Cover the hand with a glove or finger cot.

C. Cover the cut with a clean, dry bandage, and a glove or finger cot.

D. Cover the cut with a clean bandage.

10. Hands should be washed after which of the following activities?

A. Touching your hair   C. Sneezing

B. Drinking   D. All of the above

11. A foodhandler who has been diagnosed with Shigellosis should be

A. told to stay home.

B. told to wear gloves while working with food.

C. told to wash his hands every fifteen minutes.

D. assigned to a non-foodhandling position until he is feeling better.

UNIT 2

THE FLOW OF FOOD THROUGH THE OPERATION

Jack in the Box instituted HACCP in its restaurants in 1993. The program consists of farm to fork procedures from microbial meat testing by our suppliers to in-restaurant grilling procedures for ensuring properly cooked hamburgers. Jack in the Box makes proactive efforts to join forces with state legislators, regulators and advocacy groups like the National Restaurant Association Educational Foundation to help ensure safe food for customers.

David Theno, Ph.D.
Vice President, Quality Assurance, Product Safety, Research & Development
Foodmaker, Inc.

KEEP IN MIND...

The safety of the food you serve at your establishment will depend largely on your under-standing of food-safety concepts throughout the flow of food, as outlined in Chapters 5 through 8. Food safety also depends on your ability to develop a system that prioritizes, monitors, and verifies the most important food-safety practices. This system will be discussed in Chapter 9: Principles of HACCP.

*As you read through this unit, it is important to keep in mind that the food-safety concepts discussed tell you **what** to do to keep food safe, and HACCP tells you **how** to consistently keep food safe. If you know what to do, but do not know how to do it, you will not have an effective and complete food-safety system.*

Hazard Analysis Critical Control Point (HACCP) is a dynamic system that uses a combination of proper foodhandling procedures, monitoring techniques, and record keeping. The HACCP system enables you to consistently serve safe food by identifying and controlling possible hazards throughout the flow of food. Since HACCP is dynamic, it allows you to continuously improve your food-safety system. Because this is a preventive rather than reactive system, the National Restaurant Association and the Food and Drug Administration recommend that restaurants and other establishments develop and use a HACCP-based food-safety system. In fact, recognizing that the complete flow of food begins on the farm and continues through service, other government agencies and foodservice industries have accepted HACCP as the best food-safety system available. Industry segments such as canning plants, and meat, poultry, and seafood producers and processors are currently required by law to use HACCP.

Foodservice operators or managers who choose to develop a HACCP plan have found additional benefits, which can include improved quality, minimized waste, better product consistency, increased customer satisfaction, improved inspection scores, decreased customer complaints, and improved relations with the regulatory community.

Since your application of HACCP begins even before food arrives at your establishment, our presentation of the flow of food begins with pur-chasing and receiving (Chapter 5) and continues through storage (Chapter 6), preparation (Chapter 7), and service (Chapter 8). In Chapter 9, we pull the concepts from these chapters together to explain how the seven principles of HACCP are applied.

Section 5

Purchasing and Receiving Safe Food

Learning Essentials

After completing this section, you should be able to:

○ List the necessary inspection requirements, with respect to temperature, appearance, smell, and texture, for receiving the following food items:

- Seafood and shellfish
- Meat
- Poultry
- Eggs
- Dairy products
- Produce
- Refrigerated and frozen processed foods
- MAP, vacuum-packed, and *sous vide* packaged products
- Dry or canned shipments
- Hot foods

○ List the types of thermometers and outline the proper procedures for using each, including their placements.

○ Recognize that time and temperature abuse of food products can begin during the receiving process.

○ Demonstrate the proper procedures for calibrating thermometers.

TEST YOUR FOOD-SAFETY KNOWLEDGE

1. **True or False:** Upon arrival, a delivery of fresh fish should be received at 41°F (5°C) or lower. *(See Fish, page 5-6.)*

2. **True or False:** *Sous vide* (vacuum-packed) products should be received at 41°F (5°C) or lower unless specified by manufacturer's directions. *(See MAP, Vacuum-packed, and* Sous Vide *Packaged Foods, page 5-9.)*

3. **True or False:** A frozen food that is delivered in a frozen state is safe to use. *(See General Purchasing and Receiving Principles, page 5-3.)*

4. **True or False:** If a sack of flour is dry upon delivery, the contents may still be contaminated. *(See General Purchasing and Receiving Principles, page 5-3.)*

5. **True or False:** A delivery of hot roast beef and baked potatoes should be delivered at 140°F (60°C) or higher. *(See Hot Foods, page 5-10.)*

CONCEPTS

○ **Boiling-point method:** A method of calibrating thermometers based on the boiling-point of water. It can be less reliable than the ice-point method because the boiling-point of water varies with the altitude and atmospheric pressure.

○ **Calibration:** The process of making sure a thermometer gives accurate readings by adjusting it to a known standard, such as the freezing point or boiling-point of water.

○ **Ice-point method:** A method of calibrating thermometers based on the freezing point of water. It is more reliable than, and preferred over, the boiling-point method.

○ **Inspection:** The process of making sure that food deliveries meet your standards for food safety, including temperature, appearance, and packaging.

○ **Receiving:** The process of taking food delivered into your operation. It includes unloading the supplier's truck, inspecting and accepting or rejecting the items, labeling, and storing the items in a timely manner.

○ **Temperature danger zone:** The temperature range between 41°F (5°C) and 140°F (60°C) within which most bacteria grow and reproduce.

○ **Thermometer:** A device for accurately measuring the temperature of food items or the air inside a freezer or cooler. Common types are bi-metallic stemmed thermometers and digital thermometers.

○ **Time-Temperature Indicator (TTI):** A time and temperature monitoring device that is attached to a food shipment to determine if the temperature of the product has exceeded safe limits during shipment or later storage.

○ **MAP foods:** MAP stands for Modified Atmosphere Packaging. MAP is a packaging process by which air is removed from a food package and replaced with gases such as carbon dioxide and nitrogen. These gases help extend the product's shelf life.

○ **Vacuum-Packed foods:** Vacuum packaging are the process of removing air from around a food product sealed in a package. This process increases the shelf life of the product.

○ **Sous Vide foods:** Foods processed by this method are vacuum-packed in individual pouches, partially or fully cooked, and then chilled. These foods are often heated for service in the establishment.

○ **UHT foods:** These foods are heat treated at very high temperatures (pasteurized) for a short time to kill microorganisms that can cause illness. These foods are then packaged under sterile conditions. UHT foods can be stored at room temperature if unopened.

GENERAL PURCHASING AND RECEIVING PRINCIPLES

○ Buy only from suppliers who are getting their products from licensed reputable purveyors and manufacturers who inspect goods and adhere to all applicable health regulations.

○ Schedule deliveries for off-peak hours and make sure enough trained staff are available to receive, inspect, and store food promptly. Receive only one delivery at a time.

○ Carefully inspect deliveries for proper labeling, temperature, appearance, and other factors important to safety.

○ Use properly calibrated thermometers to check the temperature of received food items.

○ Check shipments for intact packaging. Broken boxes, leaky packages, or dented cans may be signs of mishandling and could be grounds for rejecting the shipment. Check packaging for signs of re-freezing, pre-wetness, and pest infestation. Simply because a product is dry or frozen upon receipt does not mean it was not wet or had not thawed during prior handling.

○ Inspect deliveries immediately and put items away as quickly as possible.

○ Label all items with the delivery date or a use-by date.

Receiving Criteria for Different Foods

Product	Accept Criteria	Reject Criteria
Meat Receive at 41°F (5°C) or lower	Beef Color: bright cherry red Lamb Color: light red Pork Color: pink lean meat, white fat Texture: firm and springs back when touched	Color: brown or greenish; brown, green, or purple blotches; white or green spots Texture: slimy, sticky, or dry Packaging: broken cartons, dirty wrappers, or torn packaging Odor: sour odor

Acceptable Beef   Unacceptable Beef

Other Criteria:

Meat must display mandatory USDA Inspection Stamps indicating product and processing plants have been inspected for sanitary standards by USDA or state department of agriculture.

Grading stamps on product indicate the level of quality and are not mandatory.

USDA
Inspection
Stamp

USDA
Grading
Stamp

Receiving Criteria for Different Foods *(Continued)*

Product	Accept Criteria	Reject Criteria
Poultry Receive at 41°F (5°C) or lower	Color: no discoloration Texture: firm and springs back when touched Odor: none	Color: purple or green discoloration around the neck; dark wing tips (red tips are acceptable) Texture: stickiness under the wings or around joints Odor: abnormal, unpleasant odor

Acceptable Poultry | Unacceptable Poultry

Other Criteria:

Fresh poultry should be received packed on self-draining crushed ice or in chill packs.

Poultry must display mandatory USDA Inspection Stamps indicating product and processing plants have been inspected by USDA or state department of agriculture.

Grading stamps on product indicate the level of quality and are not mandatory.

USDA Inspection Stamp   USDA Grading Stamp

Receiving Criteria for Different Foods *(Continued)*		
Product	**Accept Criteria**	**Reject Criteria**
Fish Receive at 41°F (5°C) or lower	Color: bright red gills; bright shiny skin Odor: mild ocean or seaweed smell Eyes: bright, clear, and full Texture: firm flesh that springs back when touched	Color: dull gray gills; dull dry skin Odor: strong fishy or ammonia smell Eyes: cloudy, red-rimmed, sunken Texture: soft, leaves an imprint when touched

Acceptable Fish   Unacceptable Fish

Other Criteria:

Fresh fish should be received packed on crushed or flaked self-draining ice

| **Shellfish**
Clams, mussels, oysters

Receive at 45°F (7°C) or lower | Odor: mild ocean or seaweed smell
Shells: closed and unbroken (indicates shellfish are alive)
Condition: if fresh, they are received alive | Odor: strong fishy smell
Shells: open and broken (indicates shellfish are dead)
Condition: dead on arrival
Texture: slimy, sticky, or dry |

Other Criteria:

Must be purchased from certified shellfish suppliers listed on Public Health Service FDA lists or lists of state-approved suppliers.

Must be received with shellstock identification tags which must remain on the original container. Operators must write the date of delivery on the tags.

Shellstock tags must be kept on file for 90 days from the date that the last shellfish was used. Different batches must not be mixed.

| **Crustacea**
Lobster, shrimp, crabs

Receive at 45°F (7°C) or lower | Odor: mild ocean or seaweed smell
Shells: hard and heavy for lobsters and crabs
Condition: if fresh, they must be received alive; packed with seaweed and kept moist | Odor: strong fishy smell
Shells: soft
Condition: dead on arrival, tail fails to curl when lobster is picked up |

Receiving Criteria for Different Foods *(Continued)*		
Product	**Accept Criteria**	**Reject Criteria**
Eggs (shell) Receive at an air temperature of 45°F (7°C) or lower	Odor: none Shells: clean and unbroken Condition: firm, high yolks that are not easy to break and whites that cling to the yolk	Odor: abnormal smell Shells: dirty or cracked

Other Criteria:

Eggs must be purchased from government-inspected suppliers and display a mandatory inspection stamp.

Eggs may display a voluntary grading stamp which indicates that they have been graded for quality under federal and state supervision.

Eggs must be delivered in trucks capable of documenting air temperature during transportation.

Eggs should be delivered within a few days of the packing date.

Once received, eggs must be stored immediately at 41°F (5°C).

Liquid, frozen, and dehydrated eggs must be pasteurized and display a USDA inspection mark.

USDA
Inspection
Stamp

USDA
Grading
Stamp

Dairy Milk, butter, cheese Receive at 41°F (5°C) or lower unless otherwise specified by law	Milk: sweetish flavor Butter: sweet flavor, uniform color, firm texture Cheese: typical flavor and texture and uniform color	Milk: sour, bitter, or moldy taste Butter: sour, bitter, or moldy taste; uneven color; soft texture Cheese: unnatural mold, uneven color, abnormal flavor or texture

Other Criteria:

Purchase pasteurized dairy products only.

Milk should be labeled Grade A, which means it meets FDA standards for quality and sanitary processing.

Dairy products with a Grade A label are made with pasteurized milk.

Receiving Criteria for Different Foods *(Continued)*		
Product	**Accept Criteria**	**Reject Criteria**
Produce Receiving temperatures vary with each produce item		Insect infestation, mold, cuts, mushiness, discoloration, wilting, dull appearance, unpleasant odors and tastes

Other Criteria:

Grounds for rejecting one produce item may not apply to another.

Cut melon is a potentially hazardous food and must be received at 41°F (5°C).

Refrigerated and Frozen Processed Foods Pre-cut meats, frozen or refrigerated entrées, fresh cut fruits and vegetables Refrigerated: receive at 41°F (5°C) or lower unless specified by the manufacturer Frozen: frozen foods should be received frozen	Package intact and in good condition	**Both Refrigerated and Frozen Products** Torn packages or packages with holes Expired use-by dates **Frozen Processed Food** Large ice crystals on the product or package (evidence of thawing and refreezing) Fluids or frozen liquids at the bottom of a case, or water stains on packaging (evidence of thawing and/or refreezing) Abnormal color Dry texture 

Receiving Criteria for Different Foods (Continued)

Product	Accept Criteria	Reject Criteria
MAP, vacuum-packed and *sous vide* packaged foods Fresh-cut produce items, bacon, some frozen diet entrées Refrigerated: receive at 41°F (5°C) or lower unless specified by the manufacturer Frozen: frozen foods should be received frozen	Package intact and in good condition	Leaking package Expired code date Unacceptable product color Product appears slimy or has bubbles

Other Criteria: Follow receiving and storage temperatures from the manufacturer if provided.

Product	Accept Criteria	Reject Criteria
Canned Goods	Can and seal are in good condition	Swollen ends, leaks and flawed seals, rust, dents, no labels

Other Criteria: Spot-check contents of a canned goods shipment. Foods that are foamy or milky should be thrown out and never tasted.

Product	Accept Criteria	Reject Criteria
Dry Goods Received at room temperatures	Package intact and in good condition	**Packaging** Holes, tears, or punctures Dampness or moisture stains (indicates that it has been wet) **Product** Contains insects, insect eggs, or rodent droppings Has an abnormal color or odor, spots of mold, or a slimy appearance

Holes and Tears

Evidence of Wetness

Receiving Criteria for Different Foods *(Continued)*		
Product	**Accept Criteria**	**Reject Criteria**
Aseptically Packaged and UHT Foods Shelf-stable milk, juice, and pudding	Package and seal intact and in good condition	Leaking, punctured, or broken packaging
Other Criteria: Stored at 41° F (5°C) when opened.		
Hot Foods Delivered at 140°F (60°C) or higher	Appropriate containers that are able to maintain temperatures	Inappropriate containers
Other Criteria: The supplier must have a HACCP plan in place or other means to verify that proper time and temperature requirements are met during the cooking process.		

MONITORING TIME AND TEMPERATURE

If food is kept in the temperature danger zone (41°F to 140°F or 5°C to 60°C) for longer than four hours, it must be discarded. That four-hour time period begins when the food is taken off the delivery truck and continues through product storage, preparation, and cooking.

Types of Thermometers

The two most common types of thermometers used in the industry are the bi-metallic stemmed thermometer and the digital thermometer.

Bi-metallic Stemmed Thermometer

○ This inexpensive thermometer is often capable of measuring temperatures from 0°F to 220°F (-18°C to 104°C). When selecting this type of thermometer, look for the following features:

- An adjustable calibration nut
- Easy-to-read temperature markings
- A dimple to mark the end of the sensing area
- Accuracy to within ±2°F or ±1°C

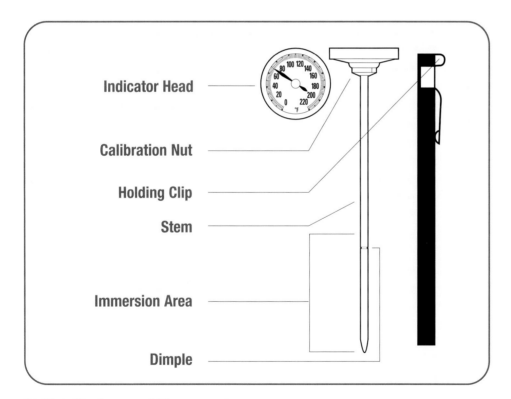

Bi-Metallic Stemmed Thermometer
Note the stem, immersion area, dimple, calibration nut, and indicator head.

Digital Thermometer

Digital thermometers measure temperatures through a metal probe or sensing area and display the results on a digital readout. They may use either a thermocouple or thermistor sensor to sense temperature. They come in a variety of styles and sizes from small pocket models to panel-mounted displays.

Many digital thermometers come with interchangeable temperature probes designed to measure the temperature of equipment and food. These include the following:

○ Immersion probe: used to measure the temperatures of liquids such as soups, sauces, or frying oil

○ Penetration probe: used to measure the internal temperature of foods such as hamburger patties, roasts, fish, etc.

○ Surface probe: used to measure the temperature of flat cooking equipment such as griddles

○ Air probe: used to measure the temperature inside walk-ins or ovens

Digital Thermometer

Thermometer Manufacturing, Atkins Technical, Gainesville, FL.

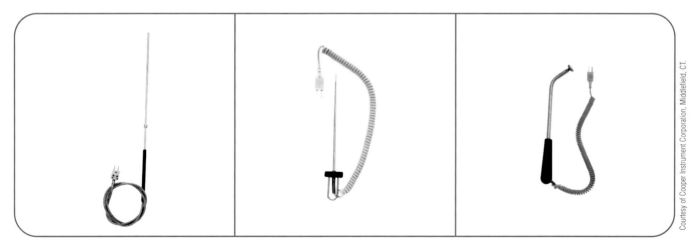

Courtesy of Cooper Instrument Corporation, Middlefield, CT.

Types of Temperature Probes
From left to right: immersion probe, penetration probe, and surface probe.

Calibrating Thermometers

Thermometers should be calibrated using either the ice-point or boiling-point method. The ice-point method is typically used unless a thermometer is not capable of registering a temperature of 32°F (0°C). The boiling-point method is sometimes less reliable due to variations in altitude and atmospheric pressure.

Ice-Point Method		
Step	**Process**	**Notes**
1	Fill a large glass with crushed ice. Add clean tap water until the glass is full.	Stir the mixture well.
2	Put the thermometer or probe stem into the ice water so that the sensing area is completely submerged. Wait 30 seconds.	Do not let the stem touch the bottom or sides of the glass. The thermometer stem or probe stem must remain in the ice water.
3	Hold the adjusting nut securely with a wrench or other tool and rotate the head of the thermometer until it reads 32°F (0°C).	Press the reset button on a digital thermometer to adjust the readout.

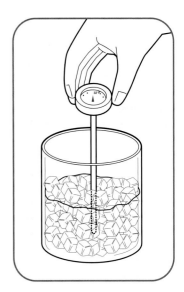

Using the Ice-Point Method

Boiling-Point Method		
Step	**Process**	**Notes**
1	Bring clean tap water to a boil in a deep pan.	
2	Put the thermometer or probe stem into the boiling water so that the sensing area is completely submerged. Wait 30 seconds.	Do not let the stem touch the bottom or sides of the pan. The thermometer stem or probe stem must remain in the boiling water.
3	Hold the adjusting nut securely with a wrench or other tool and rotate the head of the thermometer until it reads 212°F (100°C) or the appropriate boiling temperature.	The boiling point of water is about 1°F (about 0.5°C) lower for every 550 feet (168 m) you are above sea level. Press the reset button on a digital thermometer to adjust the readout.

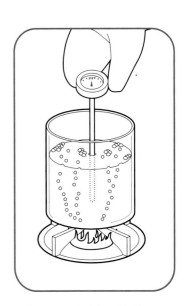

Using the Boiling-Point Method

Checking the Temperatures of Various Foods

Item	Method	Example
Meat, Poultry, Fish	Insert the thermometer or probe directly into the thickest part of the product (usually the center).	
Packaged food (refrigerated and frozen)	Insert the thermometer stem or probe between two packages, being careful not to puncture them.	
Milk and other liquids	Insert the thermometer stem or probe until at least 2 inches (5 cm) are submersed. Don't let the thermometer or probe touch the sides of the container.	
Bulk milk or liquids	Fold the bag over the stem of the thermometer or probe.	
Live shellfish	Insert the thermometer stem or probe into the middle of the carton or case, between the shellfish.	
Shucked shellfish	Insert the thermometer stem or probe into the container until the sensing area is completely submersed.	

General Thermometer Guidelines

○ **Keep thermometers and their storage cases clean.** Wash, rinse, sanitize, and air dry thermometers before and after each use to prevent cross-contamination. Keep a sufficient supply of them near the receiving area.

○ **Adjust the accuracy of all thermometers regularly.** Do this before each shift or before each day's deliveries and after they suffer a shock, such as an extreme temperature change or being dropped.

○ **Never use glass thermometers filled with mercury or spirits in your establishment.** If one should break, it can pose a serious danger to customers and employees.

○ **Wait for the thermometer reading to steady before recording the temperature.** Wait at least 15 seconds from the time the thermometer stem or probe is inserted into the food.

○ **When taking the temperature of meat, place the thermometer stem or probe into the thickest part of the meat (usually the center).** It is also a good practice to take at least two readings, which should be in different locations since product temperatures may vary across the food portion. When taking the temperature of liquids, do not let the sensing area of the stem or probe touch the sides or the bottom of the container.

Time-Temperature Indicators (TTI)

Time-Temperature Indicators (TTI) are designed to monitor both time and product temperature. Some suppliers attach these self-adhesive tags or sticks to a food shipment to determine if the temperature of the product has exceeded safe limits during shipment or later storage. If the product temperature has exceeded these limits, the TTI provides an irreversible record of the incident. A change in color inside the indicators or windows of the TTI notifies the receiver that the product has experienced time and temperature abuse. If the TTI indicates temperature abuse, the product should be rejected.

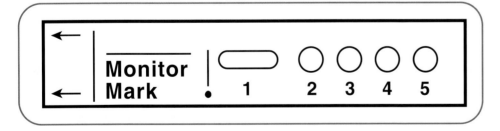

Time-Temperature Indicator
A change in color in the windows of this TTI alerts the receiver that time and temperature abuse has occurred.

SUMMARY

When receiving food deliveries, inspect the delivered items and store them promptly. The longer food spends in the temperature danger zone while you are receiving it, the less time you have to prepare it and the more likely it is to spoil. Check the temperatures of all refrigerated and frozen items with sanitized, properly calibrated thermometers. Use only reputable suppliers who follow good food-safety procedures themselves. Your business can be damaged if food is improperly handled, regardless of whether you, or someone else mishandled it before receiving it.

ACTIVITY

Crossword

Across:

1. A device for measuring the temperature of food items or the air inside a freezer or cooler.

2. The process of taking food delivered into your operation. It includes unloading the supplier's truck, inspecting the items, accepting or rejecting the items, labeling, and storing the items in a timely manner.

4. A device that is attached to a food shipment to determine if the temperature of the product has exceeded safe limits during shipment or later storage.

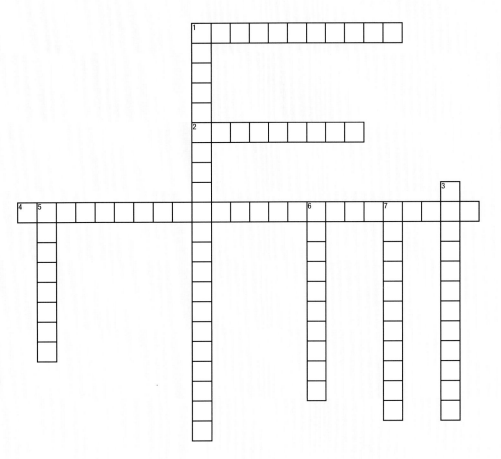

Down:

1. The temperature range between 41°F (5°C) and 140°F (60°C) within which most bacteria grow and reproduce.

3. A method of calibrating thermometers that can be less reliable because of variations in altitude and atmospheric pressure.

5. A method of calibrating thermometers that is most widely used due to its reliability.

6. The process of making sure food deliveries meet your standards for food safety, including temperature, appearance, and packaging.

7. The process of making sure a thermometer gives accurate readings by adjusting it to a known standard.

ACTIVITY

Word Find

Find the terms that go with the clues below.

Clues

1. A method of calibrating thermometers that can be less reliable because of variations in altitude and atmospheric pressure.

2. The process of making sure a thermometer gives accurate readings by adjusting it to a known standard.

3. A method of calibrating thermometers that is most widely used due to its reliability.

4. The process of making sure food deliveries meet your standards for food safety, including temperature, appearance, and packaging.

5. The process of taking food delivered into your operation. It includes unloading the supplier's truck, inspecting the items, accepting or rejecting the items, labeling, and storing the items in a timely manner.

6. The temperature range between 41°F (5°C) and 140°F (60°C) within which most bacteria grow and reproduce.

7. A device for measuring the temperature of food items or the air inside a freezer or cooler.

8. A device that is attached to a food shipment to determine if the temperature of the product has exceeded safe limits during shipment or later storage.

```
I C Y G V H V L R N M A C X I Z I R P M L X R R
C H A C F Y H K Q Q L M P A N A F L Q I N E O O
E Q T L H J W Z T W N O D E C G U M W B T T J Z
P K P E I K Z U P W C A F K V W O B A E A X X P
O T V Z M B A O N F G X Y N Q B F Q M C T Q C Z
I P Y W M P R Z R R H L A E J E G O I G D W P T
N F Z Y S D E A H Z P Q P T S M M D G U A M A S
T A I H Q C T R T D E W R Q L R N A M V J Z I L
U E Z I M E B E A I B P W Z E I Z V G J L E S R
V C N G P X M R N T O W A H E G V O S C I Y E F
Z T A U Q C O X S P U N T R L K H D H J E K C Y
J K C W Z M K F B I Z R U W W E W Z E N J C T S
U O D W I K B N R R Y T E I Q K L B Z J R M P R
B B K Y H U R Y D J A Y W D N B N V V J S J G X
Z L V G K U C M G R R F P H A S W N O A F Q V Y
C C Z P J Q R N E F K P B I Z N P E H P S T D D
T E V J M K I P P X Q N F S H R G E Y Q Q T P R
B C I K N V M L M Q F Z T M Z X X E C M V N D E
A T X T I E A T W D Y V D Y G U R S R T J H B J
N S Z E T V U F G B F R R W P Y P L T Z I X G I
O D C E Y S E R N U B R M Q N G P M K E O O J O
M E M I G D Z I S A M D X X X V O E V D Y N N U
R I C J E V R H Y R A D M O P E E G M T Q T E D
T P F B R U L X J K R M T N I O P G N I L I O B
```

ACTIVITY

What's Wrong with This Picture?

A. The picture below shows a situation involving receiving food items. How many things can you find wrong with the picture?

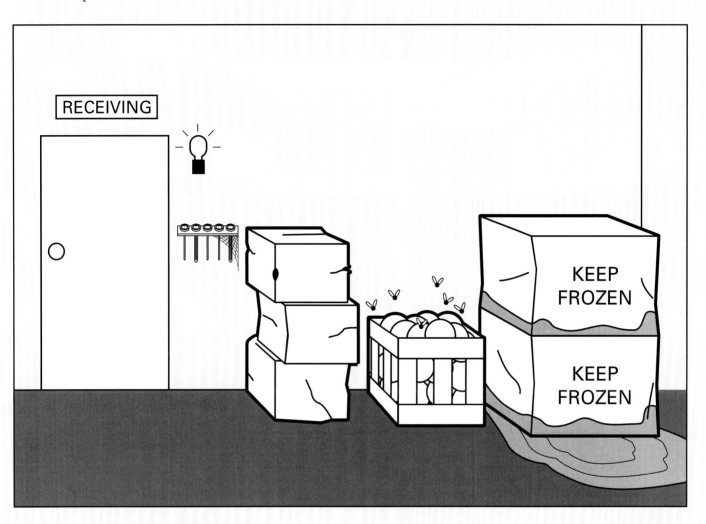

B. Considering the bad practices noted above, develop a process for receiving different foods that is consistent with the receiving guidelines noted in this section.

ACTIVITY

John's Dilemma

This exercise can be performed in pairs working independently or in front of a group. Each participant's approach to the problem should be critiqued at the end of the exercise.

Directions: Ask participants to read the case study below. Assign the role of **"John"** and **"Manager"** and give them the following guidelines:

John: Explain to the manager how you handled the receiving situation in the Case in Point. Assure him or her that everything is okay and not to worry.

Manager: Ask John specific questions about what he did, so you can determine if proper procedures were followed. If you are not satisfied with what happened, give John some guidelines on handling a similar situation in the future.

A CASE IN POINT

Case Study

ABC Seafood makes its usual Thursday afternoon delivery to The Fish House. John, a prep cook, is the only person in the kitchen when the driver rings the bell at the back door. The kitchen manager, who is in charge of receiving, is in a managers' meeting. The chef is out doing an errand, and the rest of the kitchen staff is on break, though some are still in the restaurant.

John follows the driver onto the dock where the driver unloads two crates of ice-packed fresh fish, a crate of live mussels, two crates of live oysters, a case of shucked oysters in plastic containers, and a case each of frozen shrimp and frozen lobster tails.

John goes back into the kitchen to get a stemmed pocket thermometer and remembers to look in the chef's office for a copy of the order form. He takes both out to the dock and begins to inspect the shipment.

John checks the product against both the order sheet and the invoice, then begins to check product temperatures. First he puts the thermometer probe between the packages inside the cases of frozen shrimp and lobster. Then he checks the temperature of the shucked oysters by taking the cover off one of the containers and inserting the probe into the container. The thermometer reads 45°F (7°C), which worries him. He remembers being taught that refrigerated products should be 41°F (5°C) or below. He decides not to say anything.

After wiping the thermometer probe off on his apron, John checks the internal temperature of one of the whole fish on top of one of the ice-pack crates. Finally, he reaches into one of the crates of live oysters with his hand to see if it feels cold. He notices a few mussels and oysters with open and broken shells. He removes those with broken shells, knowing the chef won't use them.

John records all his findings on the order sheet he took from the chef's office, signs for the delivery, and starts putting the products away. He puts away the live shellfish first, dumping the few that are still left in the refrigerator into the new crates so there is room. Then he puts the shucked oysters and fresh fish in the refrigerator and the shrimp and lobster into the freezer.

ACTIVITY

You're the Boss

Take approximately five minutes and draw up some receiving guidelines for your operation. You can keep it very general or add as much detail as you can within the allotted time. Be sure to cover the various aspects of receiving, inspecting, and storing the kinds of items you use in your operation. At the end of five minutes, pair up with another individual, preferably one with an identical, or at least similar, operation. Take five minutes to discuss each other's guidelines, focusing on the most effective receiving techniques as well as potential omissions.

NOTES

TRAINING TIPS

Training Tips for the Classroom

1. Thermometer Calibration Demonstration

Objective: *After completing this activity, class participants should be able to identify the two acceptable methods for calibrating a bi-metallic stemmed thermometer, and demonstrate how to calibrate one using each of these methods.*

Directions: Calibrate a bi-metallic stemmed thermometer for the class. The ice-point method is most practical, but if possible, verify the calibration using the boiling-point method after the ice-point method. Allow time for one or two members of the class to calibrate as well.

2. Temperature Instrument Group Discussion

Objective: *After completing this activity, class participants should be able to explain how each temperature instrument functions and determine which ones would be best suited to their own establishments.*

Directions: Have your class discuss the primary use, as well as the advantages and disadvantages, of the following instruments for measuring temperature:

○ bi-metallic stemmed thermometer

○ thermistor

○ thermocouples

○ various hanging thermometers

3. Receiving Game

Objective: *After completing this activity, class participants should be able to list signs of time or temperature abuse for all major food groups, and when given a specific picture or verbal description of a food product, determine if it should be accepted or rejected.*

Directions: Break the class into groups, so that you have four or five teams in the class, and give each group a bell.

Explain to the teams that you will show them a picture (or alternatively, give them a verbal description) of a product with certain characteristics (for example, a fish with sunken eyes, a dented can, discolored poultry, a properly packaged dry or canned product).

As soon as a team decides if they can accept or reject the product, they should ring their bell. They then state if they should accept or reject, giving their reasons. If correct, they receive a point. They double their points if they can state the "opposite" conditions which would make the product acceptable or unacceptable. (Give bonus points to any team that can identify a product where receiving is a Critical Control Point.)

4. Receiving Role-Play

Objective: *After completing this activity, class participants should be able to determine when to accept and reject a food product and be able to properly reject a food product from a delivery person.*

Directions: Ask two volunteers from the class to role-play a scene of a driver delivering several different types of food products to a manager in an establishment. Give the volunteers a list of specific food products that are to be delivered, and allow them five or ten minutes to prepare a role-play of how to refuse delivery of these products.

While the volunteers are preparing their scene, ask the rest of the class to each take out a blank piece of paper. When the role-play begins, members of the class are to write down as many "wrongs" in the receiving process as they see.

At the end of the role-play (after a round of applause!) ask the class how many different "wrongs" they have listed. The student with the most "wrongs" is to read his or her list. With the entire class participating, ask for the proper action that a foodservice manager or trained receiving employee should take when food products are delivered that are unacceptable.

Ask the group if anyone had additional "wrongs" that weren't discussed. Award prizes to both the role-play volunteers and the top "wrong-catchers" in class.

Training Tips on the Job

1. Field Trip to a Manufacturer or Supplier

Purpose: *After completing this activity, managers and supervisors will be able to identify the food-safety practices used by their manufacturers or suppliers to keep products safe. Note: Managers should become more knowledgeable about the products they handle and serve after this trip. It is always a good practice to go to the source and learn from the experts.*

Directions: Take your managers, supervisors, and any other interested employees on a field trip to a food manufacturer or distributor that you do business with, and get a tour of their facility.

Consider places such as a:

○ dairy plant   ○ seafood supplier

○ poultry plant   ○ bakery

○ meat-packing house   ○ food distribution warehouse

○ cannery

However, you should plan ahead. Group sizes may be limited, and access to different parts of these facilities is often restricted. (With the cooperation of a plant manager, these tours can be remarkably informative.)

2. Guest Lecturer: Thermometer Manufacturer Representative

Purpose: *After completing this activity, all staff members will be able to calibrate, properly use, and sanitize the thermometers utilized in their operation. Note: A manufacturer's rep will add credibility to the presentations, and because a new person is doing the presenting, this training may help hold the participants' attention.*

Directions: Bring in a knowledgeable and interesting manufacturer's rep from the company that manufactures the thermometers you use, and have him or her make a presentation on thermometers to your kitchen staff.

The rep should discuss the thermometers that are being used in your establishment, explaining how they are made, how they are calibrated, and how they are to be properly used and sanitized.

Make sure you are bringing someone in who is enthusiastic, is knowledgeable, and can present well. Provide information on who is in the audience, how long the presentation should be, what audiovisual equipment is available, how big and where the space is for the presentation, and so forth. Make it clear this is not to be a sales pitch.

Applications on the Job

3. Receiving Checklist

Purpose: *Having your receiving personnel work directly with suppliers to create a receiving checklist will create buy-in for both parties. Also, each group will have a better understanding of the other's needs.*

Directions: Working with both your receiving personnel and your suppliers, develop a checklist of acceptable and unacceptable conditions for the food items you normally receive.

Break down food items by category, then by specific item if necessary. Certain types of products may be listed as a general group, such as canned or dry foods. For example:

Whole fresh chicken

Acceptable:
- ○ shipped in crushed ice
- ○ temperature of 41°F (5°C)
- ○ free of discoloration
- ○ free of off-odors
- ○ free of stickiness
- ○ USDA or state agriculture stamp

Unacceptable:
- ○ shipped without ice
- ○ temperature above 41°F (5°C)
- ○ green, purple, or gray discoloration
- ○ off-odors
- ○ slime or stickiness on skin

4. Developing a Business Agreement with Your Suppliers

Purpose: *Developing a business agreement with your supplier, which identifies clearly defined standards, will help ensure that you receive safe products from them.*

Directions: Working with your receiving agents and your suppliers, develop a mutually acceptable agreement between your establishment and your food suppliers that clearly states your interest in receiving the safest food products possible.

Consider the following:

○ Work only with HACCP-based suppliers.

○ Receive deliveries at specified times and days.

○ Receive deliveries from clean, temperature-controlled trucks.

○ Receive foods at proper temperatures.

○ Receive only products that meet your checklist standards.

○ Receive foods in properly labeled, sealed packages or containers.

○ Receive foods free of pest infestation.

○ Specify that foods are to be inspected and signed for by authorized personnel only.

Having such an agreement with your suppliers helps establish a professional standard with the companies you work with, in order to help ensure a safe food product to your customers.

MULTIPLE-CHOICE STUDY QUESTIONS

1. Which is most important in choosing a food supplier?

   A. It meets your food-safety standards.

   B. Its prices are the lowest.

   C. Its warehouse is close to your establishment.

   D. It offers a convenient delivery schedule.

2. When you are receiving a delivery of food for your establishment, it is important that you

   A. refrigerate it before checking it.

   B. check it before accepting it.

   C. put it with other recent deliveries.

   D. take it out of its original packaging before storing it.

3. How might you tell by looking at it that a food delivery should be rejected?

   A. You find a cracked egg in a box of eggs.

   B. You see frost on the outside of a box of frozen corn.

   C. The ice that was packed around fresh fish has melted, and the product is above 41°F (5°C).

   D. A box of lettuce contains some field soil.

4. How should fresh salmon be packaged for delivery and storage until cooking?

   A. Layered with salt

   B. Vacuum sealed

   C. Wrapped in dry, clean cloth

   D. Covered with crushed self-draining ice

5. What is the most important observation to make about the condition of a shipment of live oysters?

   A. Their shells are closed.

   B. They smell like fresh sea water.

   C. Their shells are open.

   D. They are a uniform color and size.

6. A box of sirloin steaks carries a USDA inspection stamp and also a USDA Choice grade stamp. What do these stamps tell you?

   A. The farm that supplied the beef uses only USDA-certified animal feed.

   B. The meat processor that packaged the steaks meets USDA sanitary standards, and the meat quality is acceptable.

   C. The meat wholesaler meets USDA quality grading standards.

   D. The steaks are free of disease-causing microorganisms.

7. How could you tell if a whole fresh chicken has been exposed to temperature abuse?

   A. It came from an unlicensed supplier.

   B. The wing tips are brown.

   C. It has been in your refrigerator for more than one day.

   D. The skin is dry.

8. To check the freshness of a delivery of shell eggs, you crack one onto a plate. What about the egg would tell you that you can accept the delivery?

   A. It smells like a newly lighted matchstick.

   B. The shell cracks in half perfectly.

   C. It comes out of its shell completely.

   D. The yolk is high and firm.

9. Statements from a dairy supplier's sales brochure are listed below. Which statement should tell you not to hire this supplier?

   A. We make our cheese with only the freshest unpasteurized milk.

   B. From farm to you, our milk is kept at temperatures below 41°F (5°C).

   C. Money-back guarantee—our prices are the lowest in the area.

   D. We deliver according to your schedule and needs.

10. Which of the following do not have to be received at 41°F (5°C) or below?

   A. Beef, poultry, shellfish, and lamb

   B. Eggs, dairy products, fruits, and vegetables

   C. Ham, luncheon meats, stuffing, and duck

   D. Broth, goose, giblets, and bacon

11. Upon delivery, what is one way to tell if a frozen food product has not been properly handled?

   A. The box is water stained.

   B. The contents of the box are frozen solid.

   C. The outside of the box is warmer than 0°F (-18°C).

   D. The box is delivered in a plastic bag.

12. When should you reject the delivery of a dry or canned food product?

   A. The label on a can of peaches is tearing.

   B. The top of a can of tomatoes is bulging out.

   C. A bag of oatmeal is delivered at a temperature higher than 41°F (5°C).

   D. A box of rice does not show a USDA inspection stamp.

13. Which is not a proper step in the ice-point method of calibrating a thermometer?

   A. First, insert the probe of the thermometer into a glass of ice water.

   B. Second, wait until the temperature indicator stops moving.

   C. Third, remove the probe from the ice water.

   D. Finally, adjust the pointer on the temperature indicator to read 32°F (0°C).

14. You are the manager of a restaurant with a soup and salad bar. How should you measure the temperature of the soup to ensure that it is safe to eat?

   A. Insert an immersion probe into the center of each soup pot.

   B. Attach a surface probe to the surface of each soup pot.

   C. Ladle some soup from each pot into bowls and lay bi-metallic probes on the surface.

   D. Observe each soup pot to see if steam is rising from the surface.

15. Your manager has asked you to purchase a new thermometer for the restaurant. Which would not be a proper choice?

   A. A thermometer accurate to ±2°F (±1°C)

   B. A digital thermometer with a thermistor sensor

   C. A digital thermometer with a thermocouple sensor

   D. A mercury-filled glass thermometer

Section 6
Keeping Food Safe in Storage

Learning Essentials

After completing this section, you should be able to:

○ Explain proper storage and labeling procedures for refrigerated and frozen foods.

○ Explain the proper procedure for transferring food from its original container to another or for storing food that has been prepared or cooked, especially with respect to proper labeling.

○ Apply first-in, first-out (FIFO) practices.

○ Describe proper procedures for storing live seafood or raw food for consumption (lobster, shellfish, sushi, etc.).

○ Given a variety of food items and storage space, diagram the proper placement for each food item.

Knowledge

TEST YOUR FOOD-SAFETY KNOWLEDGE

1. **True or False:** Refrigerated, ready-to-eat foods stored at 41°F (5°C) must be eaten or thrown out within three days. *(See General Storage Principles, page 6-2.)*

2. **True or False:** Chemicals may be transferred to sturdy, properly labeled containers. *(See General Storage Principles, page 6-3.)*

3. **True or False:** Freezing destroys all harmful microorganisms in food. *(See Frozen Storage, page 6-4.)*

4. **True or False:** Melissa placed the ready-to-eat pumpkin pie directly below the raw chicken breasts. The raw chicken should have been stored on the bottom shelf below the pie. *(See Refrigerated Storage, page 6-4.)*

5. **True or False:** If stored food has passed its expiration date, you should cook and serve the food at once. *(See General Storage Principles, page 6-2.)*

CONCEPTS

○ **Refrigerated storage:** Storage used for short-term holding of fresh, perishable, and potentially hazardous food items at internal temperatures of 41°F (5°C) or lower.

○ **Deep-Chill storage:** Storage used to hold food at temperatures of 26°F to 32°F (-3°C to 0°C) for short time periods.

○ **Freezer storage:** Storage typically designed to hold food at temperatures of 0°F (-18°C).

○ **Dry storage:** The holding of nonperishable food items, such as rice, flour, crackers, and canned goods, at 50 to 60 percent humidity and between 50°F and 70°F (10°C and 21°C).

○ **First In, First Out (FIFO):** A method of stock rotation in which new supplies are shelved based on the use-by or expiration date, so that the oldest products are used first. Products with the earliest use-by or expiration dates are stored in front of products with later dates. All inventory is marked with the expiration date, when it was received, or when it was stored after preparation.

○ **Shelf life:** Recommended period of time during which a material may be stored and remain suitable for use.

○ **Hygrometer:** An instrument used to measure the humidity of a storage area.

GENERAL STORAGE PRINCIPLES

○ **Storage areas should be positioned to prevent contamination.** Foods should be stored away from warewashing areas and garbage rooms. Storage areas should be accessible to receiving, food preparation, and cooking areas to help ensure food safety.

○ **Keep potentially hazardous food out of the temperature danger zone, 41°F to 140°F (5°C to 60°C).**

○ **Follow FIFO: first in, first out.** On each package, write the expiration date, the date the product was received, or the date the product was stored after preparation. Shelve food based upon use-by or expiration dates, so the older food is used first. Regularly check expiration dates and discard food that has exceeded its expiration date.

○ **Potentially hazardous ready-to-eat foods should be discarded if not used within seven days of preparation.** Potentially hazardous, ready-to-eat foods that have been frozen should be discarded if not consumed within twenty-four hours of being thawed.

○ **Check the temperature of stored foods and storage areas regularly.** Use a calibrated thermometer to check temperatures at least once per shift.

○ **Store food only in designated storage areas.** Never store chemicals or cleaning supplies in food storage or preparation areas.

○ **Keep storage areas clean and dry.**

○ **Keep all goods in clean, undamaged wrappers, packages, or containers that are labeled with date opened or received, contents, and expiration date.**

○ **Clean carts or other vehicles that transport food.**

○ **Transfer food between containers properly.** Use leak-proof, pest-proof, nonabsorbent, sanitary containers with tight-fitting lids. Label the new container with the date and time prepared. Never use empty food containers to store chemicals or put food in empty chemical containers. If it is necessary to transfer chemicals, store them in sturdy containers clearly labeled with the contents and their hazards.

Use the FIFO Method of Storage
Shelve food according to its use-by or expiration date.

Courtesy of Daydots Foodservice Products, Fort Worth, TX, 800-321-3687.

Refrigerated Storage

The following guidelines should be followed when you are storing foods in refrigerators.

○ **Monitor food temperature regularly.** Using a calibrated thermometer, take random temperatures of food stored in the refrigerator and check the temperature of the refrigeration unit. Recalibrate unit thermometers or replace them if necessary. Hanging thermometers should be placed in the back of the refrigerator and near the door.

○ **Don't overload the refrigerator.** Overloading may prevent airflow and make the unit work harder to stay cold. Don't line the shelving with foil or paper, either. This will also prevent airflow.

○ **Use caution when cooling hot food in the refrigerator.** This could warm up the interior enough to put foods in the temperature danger zone. Keeping the refrigerator door closed as much as possible will also help keep the interior cold.

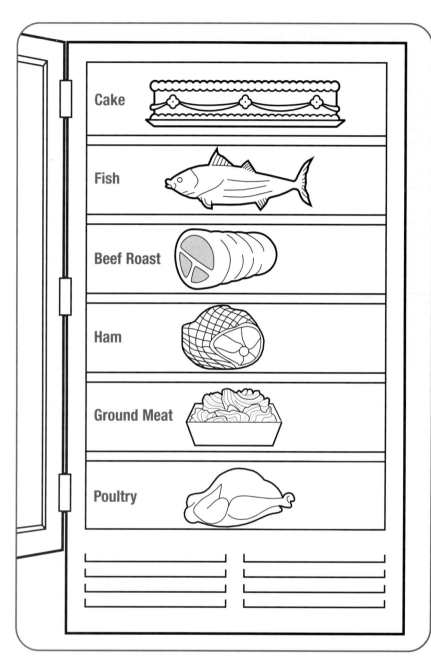

Proper Refrigerator Storage
Top-to-bottom storage of different raw and ready-to-eat products in the same refrigerator.

○ **Store raw meat, poultry, and fish separately from cooked and ready-to-eat foods whenever possible to prevent cross-contamination.** If not, always store prepared or ready-to-eat foods above raw meat, poultry and fish. Raw meat, poultry, and fish should be stored in the following top-to-bottom order in the refrigerator: fish; whole cuts of beef; pork; ham, bacon, and sausage; ground beef and ground pork; poultry.

○ **To keep food at a specific temperature, the air temperature in the refrigerator usually must be about 2°F (1°C) lower.** For example, to hold chicken at 41°F (5°C) the air temperature must be 39°F (4°C).

○ **Wrap foods properly. Leaving food uncovered can cause cross-contamination.** Products should be stored in clean, covered containers that are clearly marked.

Freezer Storage

Frozen food should be stored at temperatures that will keep it frozen. Freezing does not kill all microorganisms. However, it does slow their growth substantially. When storing food in freezers, you should follow these guidelines.

○ **Check unit and food temperatures regularly.**

○ **Rotate frozen food using the FIFO method. Check use-by dates.**

○ **Store foods in their original containers or wrap them tightly in moisture-proof containers. Clearly label containers with the contents, delivery date, and/or use-by-date.**

○ **Use caution when placing hot food in the freezer.** Warm foods may raise the temperature inside the unit and partially thaw contents.

○ **Regularly check foods that may be damaged by lengthy freezing.**

○ **Never refreeze thawed food until it has been thoroughly cooked.** Thawed food is more likely to support the growth of microorganisms.

○ **Keep the unit closed as much as possible.**

○ **Defrost freezers regularly. Move food to another freezer while defrosting.**

Deep-Chill Storage

Deep chilling is best for meat, fish, poultry, and *sous vide* foods. Store food at 26°F to 32°F (-3°C to 0°C).

Dry Storage

You should follow these guidelines when storing foods in dry storage.

○ **Keep storerooms cool, dry, and well ventilated.** Moisture and heat are the biggest dangers to dry and canned foods. The temperature of the storeroom should be between 50°F and 70°F (10°C to 21°C). Keep relative humidity at 50 to 60 percent if possible.

○ **Store foods in their original packages if possible.** Once the packages are opened, store the product in airtight containers that are clearly labeled.

○ **Store dry foods at least six inches off the floor and out of direct sunlight.**

Storing Foods Properly
Products should be stored wrapped in clean containers that are clearly marked.

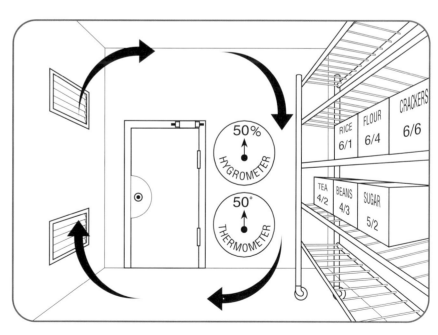

An Acceptable Dry-Storage Facility
Dry-storage temperatures should be between 50°F to 70°F (10°C to 21°C), and the humidity should be between 50 and 60 percent if possible.

Recommended Requirements for Storing Specific Foods		
Product	**Storage Temperature**	**Other Requirements**
Meat	Store fresh at 41°F (5°C) or lower	Wrap meat in airtight, moisture-proof materials.
Poultry	Store fresh at 41°F (5°C) or lower	Store ice-packed poultry as is in self-draining containers. Change the ice often and sanitize the container regularly.
		Whole birds can be loosely wrapped and should be used in 3 to 4 days.
		Fresh refrigerated parts should be used within 1 to 2 days.
Fish	Store fresh at 41°F (5°C) or lower	Store ice-packed fish as is in self-draining containers. Change the ice often and sanitize the container regularly. Fish can be stored this way up to 3 days.
		Keep fillets and steaks in original packaging or tightly wrapped in moisture-proof wrappings.
		Store frozen fish in moisture-proof wrapping.
		Fish meant to be eaten raw (except oysters and certain tuna species) must be frozen to the following temperatures before being served:
		-4°F (-20°C) or lower for 7 days (168 hours) -31°F (-35°C) or lower for 15 hours in a blast freezer.
Shellfish and Crustacea	Store live at 45°F (7°C) or lower	Store alive in the original container.
		Molluscan shellfish (clams, oysters, mussels) can be stored in a display tank if a variance is obtained from the local health department.
		Shellstock tags must be kept on file for 90 days from the date the last shellfish was used.
Eggs (shell)	Store fresh at 41°F (5°C) or lower	Keep eggs in storage until immediately before use.
	Maintain constant temperature and humidity.	Use all eggs within a few weeks of purchase.
		Liquid egg products should be stored in their original containers at 41°F (5°C) or lower. Do not freeze.
		Dried egg products can be stored in a dry, cool storeroom but should be refrigerated at 41°F (5°C) or lower when reconstituted.
Dairy	Store fresh: 41°F (5°C) or lower	Keep dairy products tightly covered and store them away from foods with strong odors.
Ice Cream and Frozen Yogurt	Frozen: 6°F to 10°F (-14°C to -12°C)	Use the FIFO method of storage. Discard product if the expiration or use-by date has passed.
Fresh Produce	Storage temperatures vary depending upon the product.	Whole raw produce and raw cut vegetables delivered packed on ice can be stored that way.
		Produce should not be washed before storage.

Recommended Requirements for Storing Specific Foods *(Continued)*		
Product	**Storage Temperature**	**Other Requirements**
MAP, Vacuum-Packed, and *Sous Vide* Packaged Foods	Store at temperatures recommended by the manufacturer or at 41°F (5°C) or lower	Discard product if the expiration or use-by date has passed. Discard product if package is torn, or if the contents are slimy or have bubbles (possible growth of *clostridium botulinum*).
Aseptic and UHT Foods	Store product at room temperature	Aseptic and UHT Foods should be stored at 41°F (5°C) or lower when opened.
Canned and Dry Foods	50°F to 70°F (10°C to 21°C)	Keep storerooms dry and the humidity low. Flour, cereal, and grain products should be stored in airtight containers. If dry foods are removed from original packaging, store them in clearly labeled containers. Check packages for insect or rodent damage.

SUMMARY

The key to keeping food safe during storage is to keep it out of the temperature danger zone where harmful microorganisms can grow and reproduce. Most unfrozen, perishable food must be refrigerated at an internal temperature of 41°F (5°C) or lower. Frozen food should be stored at temperatures that will keep it frozen. Monitor the temperature of food in refrigerated and frozen storage areas regularly. Use caution when storing hot foods in refrigerators or freezers. Doing so can raise the temperature inside the unit, which in turn could raise the internal temperature of food that has been stored. Don't overload storage units, either. This will hinder the airflow inside, reducing the unit's ability to maintain the required temperatures.

To reduce the risk of cross-contamination, never store cooked or ready-to-eat food below raw food. Raw meat, poultry, and fish should be stored in the following top-to-bottom order: fish; whole cuts of beef; pork; ham, bacon and sausage; ground beef and ground pork; poultry.

Store food in leak-proof, pest-proof, nonabsorbent, sanitary containers with tight-fitting lids. Practice FIFO. Clearly label all items with the product's expiration date, the date the product was received, or the date it was stored after preparation. Shelve new supplies based on the use-by or expiration date, and use the oldest items first. Discard food that has passed the expiration date.

NOTES

ACTIVITY

Crossword Puzzle

Across:

2. Type of storage used to keep food at an internal temperature of 41°F (5°C).

4. A recommended period of time during which a material may be stored and remain suitable for use.

8. The range where most bacteria reproduce and grow.

Down:

1. Type of storage used to keep certain types of food at 26°F to 32°F (-3°C to 0°C).

3. Can result from storing raw food above cooked or ready-to-eat food.

5. Type of storage used to keep frozen foods frozen.

6. Method of stocking in which oldest supplies are used first.

7. An instrument used to measure the relative humidity in a storage area.

9. Type of storage used to keep food at 50°F to 70°F (10°C to 21°C) and 50 to 60 percent humidity.

ACTIVITY

Word Find

Find the terms that go with the clues below.

Clues

1. The range where most bacteria reproduce and grow.

2. Type of storage used to keep food at an internal temperature of 41°F (5°C).

3. Type of storage used to keep certain types of food at 26°F to 32°F (-3°C to 0°C).

4. Type of storage used to keep frozen foods frozen.

5. Type of storage used to keep food at 50°F to 70°F (10°C to 21°C) and 50 to 60 percent humidity.

6. Method of stocking in which oldest supplies are used first.

7. Can result from storing raw food above cooked or ready-to-eat food.

```
Z S B M L E L G Q H M J R U G H A S V H I L L F
P Y K M T P Q L F T O F N D Y B S R Y E Q N S I
E A N D A S Y K M G M H B V V U M G G Y N L H R
T F Z M U K F N N Y L J W F N W R A I B Z M A S
B E I M O J N J R D V D F D D O R E D Z J A C T
Z P M L S Z S D X J Q Z V P M O R H D S R M L I
Z E P P F T D V A K X O D E T C A Q G A Z N F N
Y R Y E E L H T C M I X T S L W J S L J I O P F
E K R T S R E X Y C H E D Q F S S A N T K I E I
J O O B E J A H J X R E I I Y F P R I N E T G R
G M U C F K N T S U T N P J W R P D S B G A A S
K J Y X Z X T B U A B X C Y W W M Y O K A N R T
X I Q J U C Y G R R D F S Y U L W R K G R I O O
Y N Z E P P X E M D E R W F T W R K Q A O M T U
O N V Y S W G M I V U D Y L N R T X M C T A S T
H V X M P I A V P H D P A S E Z P G S C S T L Y
N H R Z R I N D V Q L D S N T V S G U E R N L T
U L G F X F U O M B O K L R G O G P S W E O I C
J M E X R I C J L I H P Q H N E R U R V Z C H N
Q R T T F R B A Y J U R E M X J R A U X E S C B
A D L W X Q U Z W Y Y O B W H D N Z G V E S P Y
V B H E X Y Z W X Q V Y R U U H I B O E R O E C
Y W M U M A B X Y Y M M U K A Q Z J T N F R E N
V X C Z B X B Z A X T N V P F E U S M P E C D Y
```

8. An instrument used to measure the relative humidity in a storage area.

9. A recommended period of time during which a material may be stored and remain suitable for use.

ACTIVITY

Load the Refrigerator

Draw lines from the different food items to their proper refrigerator shelves for best storage.

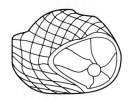

Raw pork, ham, bacon, or sausage

Raw beef

Cooked or ready-to-eat food

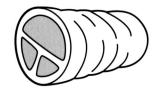

Raw chicken

Raw fish

Raw ground beef or pork

ACTIVITY

What's Wrong with This Picture?

How many problems related to food storage safety can you find?

A CASE IN POINT I

Case Study

On Monday afternoon, the kitchen staff at the Sunnydale nursing home was busy cleaning up from lunch and preparing dinner. Pete, a kitchen assistant, put a large stockpot of hot, leftover vegetable soup in the refrigerator to chill. Angie, a cook, began deboning chicken breasts that were stored earlier. When she was finished, she put the chicken on an uncovered sheet pan and stored it in the refrigerator. She carefully placed the raw chicken on the top shelf away from the hot soup. Next, Angie iced a carrot cake she had baked that morning. She put the carrot cake in the refrigerator on the shelf directly below the chicken breasts.

What storage errors were made? What food items are at risk?

A CASE IN POINT II

Case Study

Anticipating a slow dinner, Ed, the owner of Big City Diner, gave his kitchen manager the afternoon off. Some time later, Acme Distributors delivered an order of canned and dry foods and supplies, including canned tomato sauce, canned soup, crackers, pasta, paper napkins, and cleaning supplies such as warewashing detergent and sanitizing solution.

After checking the order, Ed stacked the canned and dry foods onto a dolly and wheeled them into the storeroom. There was an open case of tomato sauce on the shelf with one can left in it and a full case behind it. Ed removed the open case and replaced it with the case on the dolly. He put the single can in an open spot on another shelf. By shifting some boxes around, Ed was able to put the cases of soup, napkins, and pasta on shelves in front of cases already stored there. While moving cases, he accidentally knocked over a container of flour, spilling some on the floor. There was no room on the shelves for the crackers, so he left the case on the floor. He was sweating from exertion, and wondered if the storeroom was too hot. He checked the hanging thermometer, which read 70°F (21°C).

Ed went back to the dock to get the cleaning supplies. He wheeled those into the dish room and slid the cases under the warewashing machine. When he returned the dolly to the back door, a truck from Mike's Produce pulled up to deliver a case of lettuce, two cases of tomatoes, a case of onions, and a case of cabbage. Ed inspected the delivery and signed for it. He stacked it all on the dolly and wheeled it toward the kitchen. Suddenly remembering some paperwork he had to finish, he left the produce in the hallway outside the restroom. He knew the dinner cook would be coming in soon and would put it away.

What did Ed do wrong? What are the possible consequences?

NOTES

TRAINING TIPS

Training Tips for the Classroom

1. *"Rooms for Rent" Group Activity*

Objective: *After completing this activity, class participants will be able to identify the most effective storage procedures for freezers, refrigerators, deep-chill units, dry storage rooms, and chemical supply rooms. From this activity they will also be able to identify general principles for proper storage. Note: Class participants will meet this objective through group involvement and discussion activity, as opposed to instructor lecture.*

Directions: Create five groups in the classroom, and assign each group one of the following five storage areas:

- freezer
- deep-chill unit
- refrigerator
- dry-storage room
- chemical and cleaning supplies storage

Give each group five minutes to come up with as many storage procedures as possible for its assigned area.

When the time has expired, have each group present its list, soliciting feedback or comments from the entire class on each storage principle or guideline as it is listed.

All principles or guidelines that the class agrees are valid are recorded by a team member on a flipchart or blackboard in the front of the classroom. After a group presents its principles, solicit additional principles or guidelines from the class.

After all five groups have presented their principles, summarize the activity by making a list of general storage principles from guidelines that were common to all five storage lists.

2. *"Where Do I Go?" Team Contest*

Objective: *After completing this activity,, class participants will be able to identify proper storage areas and handling guidelines for specific types of food. Note: This is a team-building exercise.*

Directions: Divide a sheet of paper into three columns with the following headings:

Item: **Location:**   **Special instructions:**

In the *item* column, list twenty-five to thirty specific food products that may be received into a foodservice facility. Be sure to include a wide variety of foods, including both fresh and frozen meats, poultry, fish, seafood, shellfish, dairy products, eggs, MAP and *sous vide* foods, fresh produce, UHT products, dry foods, and canned or bottled foods.

Assign teams of two to four people per team, and give each team a copy of the worksheet. Give the teams five minutes to determine the proper storage area for each item, as well as at least one specific storage instruction for each item.

Then, allowing each team to score its own worksheet, go down the list of foods, coming to agreement among the class as to where the product should be stored and what specific storage instructions should be given for each item.

The team with the most correct locations and instructions wins.

3. Food Products in Storage Discussion

Objective: *After completing this activity, class participants will be able to develop standard operating procedures (SOPs) for storage of various types of food. Note: This higher-level activity may not be suitable for all classes.*

Directions: Give your class a challenge: develop SOPs for storage of the following items:

- live oysters
- MAP chicken breasts
- prepared shrimp salad
- whole red snapper
- fresh hamburger patties
- fresh shell eggs
- *sous vide* chicken cordon bleu

Divide the class into groups, and ask them to address the following questions:

1. What is the ideal temperature in storage for each product?

2. Who stores the food, where is it to be stored, and how should it be stored?

3. What documentation should you keep?

Allow ample time for each group to develop SOPs for these products in storage. Then review with the class.

Training Tips on the Job

1. "What's This?"

Purpose: *To illustrate the importance of keeping products in their original containers, labeling products in secondary containers, and storing cleaning supplies and chemicals away from food. Note: This is an easy, yet memorable, demonstration.*

Directions: Place a number of food and non-food products in unlabeled clear containers, or otherwise conceal their identity. Products should include items that are difficult to identify or distinguish from other products, such as:

- sugar
- salt
- flour
- cleanser
- baking soda
- bleach
- vinegar
- oil
- caustic solvent
- window cleaner
- pan spray
- stainless-steel cleaner

Have kitchen employees try to identify each product without smelling, touching, and certainly without tasting!

2. "What's Wrong with This Picture?" Contest

Purpose: *To involve your kitchen staff in the assessment of storage practices in your establishment and to achieve their buy-in when implementing corrective actions that result from their assessment. Note: The competitive nature of this activity should make it fun for your staff.*

Directions: Have each member of your kitchen staff conduct an inspection of the establishment's food storage areas to find mistakes in your freezers, refrigerators, and dry-storage areas, as well as in any temporary storage units for food. The inspector who finds the most mistakes wins a prize.

Equip your employees with clipboards, pencils, blank paper, and thermometers, and have them make note of everything they find to be in violation of safe food-storage principles. They should report on the condition of the storage areas and of the food itself.

When your kitchen staff has finished inspecting, collect and make copies of the reports.

Hold a group meeting, and encourage employees to share their observations, area by area. Have someone record the collective observations. Employees can give themselves a point for each noted violation. (Let them know you have copies of their reports, lest they be tempted to inflate their grades!) The staff member with the most points wins.

Let your staff know that the real winner is the entire operation, if everyone learns from the process. Let them know that the inspection is just the assessment step. Corrective actions need to be taken, and better storage procedures need to be in place. These corrective actions will maximize food safety and quality. They also help control food cost by minimizing waste. Be sure to set an action plan in place, getting everyone involved. Follow-up is essential. Let your staff know that you will be doing the next inspection. If you give storage an A+, you'll treat them all to a party!

3. Developing Charts and Checklists

Purpose: *To involve your kitchen management team in the development of storage job aids that can be used as easy reference tools for employees and management. Note: These job aids will help support your HACCP systems and will lead to management buy-in because the managers were given the opportunity to create them.*

Directions: Get your kitchen management involved in developing a set of colorful, readable, usable charts and checklists that establish your storage guidelines and procedures.

These reference tools may include:

○ A storage temperature list for foods in three areas: freezer, refrigerator, and dry storage.

○ Storage guidelines for each storage area.

○ Time-temperature logs for checking food temperatures during each shift (with corrective action listed).

○ Use-by or discard dates for food products.

Important: Be sure that the standards meet or exceed codes of your regulatory agency and any recommendations from manufacturers or processors.

Here are some general guidelines for kitchen charts or logs.

○ Keep them simple. Less is more.

○ Make them colorful. Better yet, color-code for storage area. Use red for CCPs.

○ Print them in several languages, if necessary.

○ Laminate all materials.

○ Post them in the appropriate areas.

○ Use pictures or graphics, and fewer words, whenever possible.

MULTIPLE-CHOICE STUDY QUESTIONS

1. At what storage temperature would ground beef most likely become unsafe to use?

   A. 0°F (-17°C)   C. 41°F (5°C)

   B. 30°F (-1°C)   D. 60°F (16°C)

2. Which storeroom condition is most likely to cause a sack of flour to become contaminated?

   A. A damp floor   C. An air temperature of 50°F (10°C)

   B. Fluorescent lighting   D. Full shelves

3. Under which condition could you display to customers the live mussels that you will be cooking and serving to them?

   A. You have special permission from the health department.

   B. You have a self-cleaning filtration system for the display tank.

   C. You will also cook and serve other mussels that have not been on display.

   D. You have removed the shellstock identification tags as required by law.

4. It is important that storage areas be

   A. close to where cleaning supplies are accessible.

   B. kept safe from all sources of contamination.

   C. kept warm and dry.

   D. opened only by the manager.

5. There is a date written on each of the single-serve boxes of ready-to-eat breakfast cereals in your storage room. What does this date tell you?

   A. You should manage your stock to serve the boxes within a week of the dates.

   B. The contents of the boxes will be unsafe to eat one week before the dates.

   C. You should rotate your stock to serve the boxes with the older dates first.

   D. The manufacturer will take boxes back for credit after the dates.

6. A major problem with the storage of vacuum-packaged foods is that
   A. loss of moisture can lessen food quality.
   B. heat can cause vacuum-packaged cans to explode.
   C. air leakage can contaminate neighboring foods.
   D. bacteria that cause botulism can develop.

7. If you see that a stored *sous vide* package is torn, you should immediately
   A. cook and serve the contents.
   B. throw away the contents.
   C. mend the tear and freeze the package.
   D. measure the internal temperature of the package.

8. In order to guarantee that foods in a refrigerator are being kept at 41°F (5°C) or lower, maintain the air temperature in the refrigerator at
   A. 26°F (-3°C).
   B. 32°F (0°C).
   C. 39°F (4°C).
   D. 0°F (-18°C).

9. The hanging thermometer in your refrigerator has fallen on the floor. What action should you take?
   A. Purchase a new thermometer.
   B. Rehang the thermometer in a safer place in the refrigerator.
   C. Replace the thermometer with a bi-metallic probe thermometer.
   D. Check the accuracy of the thermometer and recalibrate or replace if necessary.

10. Live oysters that will be served raw to customers must be
    A. kept at temperatures between 35°F (2°C) and 45°F (7°C) before serving.
    B. shown live to customers before serving.
    C. cooled to 32°F (0°C) right before serving.
    D. warmed to 41°F (5°C) right before serving.

Section 7
Protecting Food During Preparation

TEST YOUR FOOD-SAFETY KNOWLEDGE

1. **True or False:** After the loose dirt has been brushed off the outside of a melon, it can be safely cut open. *(See Fruits and Vegetables, page 7-6.)*

2. **True or False:** A fruit salad containing watermelon, grapes, and strawberries does not need to be maintained under 41°F (5°C) or below to be safe to eat. *(See Fruits and Vegetables, page 7-6.)*

3. **True or False:** Two-stage cooling involves first refrigerating and then freezing foods. *(See Concepts, page 7-2.)*

4. **True or False:** A casserole has been removed from the oven before it has reached a safe minimum internal temperature. Before resuming the cooking, it should first be cooled to below 41°F (5°C). *(See Poultry, page 7-8.)*

5. **True or False:** The minimum safe internal temperatures for a beef roast and a pork roast are the same. *(See Beef and Pork Roasts, page 7-8.)*

Learning Essentials

After completing this section, you should be able to:

○ Identify considerations of time, temperature, cross-contamination and personal hygiene when preparing raw and cooked foods.

○ Identify the four acceptable methods for thawing foods.

○ List the minimum internal cooking temperatures for meat, seafood, and poultry; stuffed foods; casseroles; and combined foods containing cooked and uncooked foods.

○ Identify the proper procedures for preparing raw fruits and vegetables.

○ Identify the time and temperature requirements for cooling cooked foods.

○ Identify acceptable methods for cooling foods.

○ Identify the time and temperature requirements for reheating food.

○ Identify the requirements for cooking or reheating food in a microwave including temperature, time, and stirring requirements.

○ Identify proper procedures for labeling and storing cooked foods.

CONCEPTS

○ **Temperature abuse:** Temperature abuse has occurred any time potentially hazardous food is exposed to the temperature danger zone of 41°F to 140°F (5°C to 60°C). Foods being prepared, cooked, cooled, and reheated should be passed through the temperature danger zone as quickly as possible to avoid the growth of disease-causing microorganisms.

○ **Four-hour rule:** Potentially hazardous foods may not be exposed to the temperature danger zone for more than four hours. The exposure time accumulates during each stage of handling from the time food arrives at the receiving dock to the time it is cooked. Exposure time begins again when the food is held for service, cooled, and reheated.

○ **Minimum internal temperature:** The required cooking temperature that the internal portion of food must meet in order to sufficiently reduce the number of microorganisms that might be present. This temperature is specific to the type of food being cooked. Food must reach and hold at its minimum internal temperature for a specified amount of time.

○ **Two-stage cooling method:** By this method, cooked foods must be cooled from 140°F to 70°F (60°C to 21°C) within two hours and from 70°F (21°C) to below 41°F (5°C) in an additional four hours.

○ **One-stage cooling method:** By this method, cooked foods must be cooled to 41°F (5°C) or lower within four hours.

○ **Ice-water bath:** A method of cooling food in which a container holding hot food is placed inside a larger container of ice water. The ice water surrounding the hot food container disperses the heat quickly.

○ **Cold paddle:** A plastic wand or paddle that is filled with water and frozen. When used to stir hot foods, it cools the food quickly.

○ **Pooled eggs:** The practice of combining several cracked eggs in a common container.

INTRODUCTION

The two leading factors in foodborne illness are temperature abuse and cross-contamination. The key to serving safe food is to handle food safely. Train your employees in the principles of time, temperature, and sanitation. Implement procedures to ensure safe food-handling.

Time and Temperature Control

Follow these guidelines to prevent food from being subjected to time and temperature abuse.

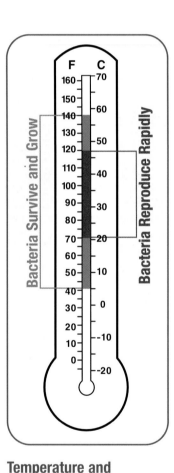

Temperature and Bacterial Growth
Most microorganisms grow rapidly at temperatures between 70°F and 120°F (21°C and 49°C).

- **Make thermometers available in the areas where they are needed and train employees to use them properly.** Record temperatures and the times they are taken on simple forms attached to clipboards near equipment.

- **Establish clear procedures.** Build time and temperature controls into each of your recipes.

- **When preparing food, take out only as much food from storage as you can prepare at one time.**

- **Cook, hold, cool, and reheat food properly.** (Requirements for each process will be covered later.)

- **When heating or cooling food, pass it through the middle of the temperature danger zone (70°F to 120°F [21°C to 49°C]) as quickly as possible.** Microorganisms grow faster in the middle of this range.

- **Discard food if it spends more than four hours total in the temperature danger zone** (41°F to 140°F [5°C to 60°C]). This includes time spent in the temperature danger zone during receiving, storage, preparation and cooking, and then again during holding, cooling, and reheating.

- **Take corrective actions if time and temperature standards are not met.**

Preventing Cross-Contamination

Guidelines for preventing cross-contamination:

- **Prepare raw meats, fish, and poultry in separate areas from produce or cooked and ready-to-eat foods.** If space is not available, prepare these items at different times.

- **Assign specific equipment (cutting boards, utensils, and containers) to each type of food product.** For example, use one set of cutting boards, utensils, and containers just for poultry, another set for meat, and a third set for produce. Color-coded cutting boards and utensils can also be assigned to a specific product.

- **Clean and sanitize all work surfaces, equipment, and utensils**

Manufactured by KatchAll Industries

Color-Coded Cutting Board
Different colored cutting boards can be designated for different foods, reducing the risk of cross-contamination.

after each task. *Note: Cleaning and sanitizing procedures will be discussed in more depth in Unit 3.*

○ **Make sure cloths or towels used for wiping spills are not used for any other purpose.** Use disposable cloths or towels, or use different (color-coded) cloths and towels for each prep area or task. Rinse the cloth or towel after each task and store it in a sanitizing solution.

○ **Make sure employees wash their hands between tasks.** Hands should be washed before a new task is started when handling raw food.

○ **Consider using single-use disposable gloves when preparing or serving food.** Employees must wash their hands before putting on gloves. Gloves should be used only for a specific task and changed each time a new task is started. If punctured or ripped, the gloves must be changed.

THAWING FOODS PROPERLY

Freezing does not kill all microorganisms, but it does slow their growth. When frozen food is thawed and exposed to the temperature danger zone, microorganisms that are present will begin to grow and multiply.

There are only four acceptable ways to thaw food.

○ **In a refrigerator, at temperatures of 41°F (5°C) or lower.**

○ **Submerged under running potable water, at a temperature of 70°F (21°C) or lower.** Water flow must be strong enough to wash loose food particles into the overflow drain.

○ **In a microwave oven, if the food will be cooked immediately after thawing.**

○ **As part of a cooking procedure, as long as the product meets the required minimum internal cooking temperature.**

PREPARING FOODS

Following the basic principles of time and temperature control, and putting up barriers to cross-contamination, will help prevent most cases of foodborne illness. However, some foods and types of preparation require a bit more care.

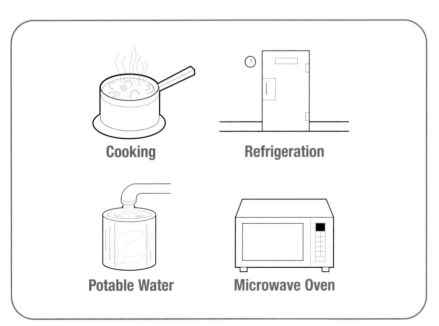

Cooking   Refrigeration

Potable Water   Microwave Oven

Acceptable Methods of Thawing Food

Meat, Fish, and Poultry

The source of most cross-contamination in an operation is raw meat, poultry, and seafood. Aside from the basic cross-contamination guidelines mentioned earlier, follow these guidelines:

○ Use clean and sanitized work areas, cutting boards, knives and utensils.

○ Take out of storage only as much product as you can prepare at one time.

○ Put raw prepared meats back into storage as quickly as possible, or cook them as soon as possible. Store them properly to prevent cross-contamination.

Protein Salads

Chicken, tuna, egg, pasta, and potato salads have been known to cause outbreaks of foodborne illness. Since these foods are not typically cooked after preparation, there is no chance to kill microorganisms that may have been introduced during preparation. Therefore, extreme care must be taken when preparing protein salads. Follow these preparation guidelines:

○ **Make sure that proteins for salads (eggs, chicken, tuna, and other meats) have been properly cooked, held, cooled, and stored.**

○ **Make sure leftovers used for salads have not been left in the refrigerator too long.** Cooked leftover meat must be discarded after seven days.

○ **Leave food in the refrigerator until all items are ready to be mixed.**

○ **Consider chilling all ingredients before making salads.** For example, tuna and mayonnaise can be chilled before the tuna salad is made.

○ **Prepare small batches, so large amounts of food don't sit out at room temperature for long periods of time.**

Eggs and Egg Mixtures

Eggs are considered to be a potentially hazardous food because they are able to support the rapid growth of microorganisms. When preparing eggs and egg mixtures, follow these guidelines:

○ **Handle pooled eggs (if allowed) with special care.** Pooled eggs are eggs that are cracked open and combined in a common container. They must be handled with care because bacteria in one egg can be spread to the rest. Pooled eggs must be cooked promptly after mixing, or stored at 41°F (5°C) or lower for no more than two hours. Containers that have held pooled eggs must be washed and sanitized before being used for a new batch.

○ **Operations that serve highly susceptible populations, such as hospitals and nursing homes, must always use pasteurized eggs in recipes that require eggs.**

○ **Promptly clean and sanitize all equipment and utensils that were used when preparing eggs.**

○ **Consider using pasteurized eggs when preparing egg dishes that require little or no cooking.** These include dishes like mayonnaise, eggnog, Caesar salad dressing, and hollandaise sauce.

Batters and Breading

Batters and breading can be hazardous if made with milk or eggs. These products face the same dangers from temperature abuse and cross-contamination as other foods, so they should be handled with care. If you make breaded or battered foods from scratch, follow these guidelines:

○ **Prepare batters in small batches.** Store what you don't need at 41°F (5°C) or lower in a covered container.

○ **When breading food that will be cooked at a later time, store the breaded food in the refrigerator as quickly as possible.**

○ **Throw out any unused batter or breading after each shift.** Never use batter or breading for more than one product.

○ **Make batters with pasteurized egg products whenever possible.**

Fruits and Vegetables

When preparing fruits and vegetables, follow these guidelines:

○ **Make sure fruits and vegetables do not come in contact with surfaces that have been exposed to raw meat and poultry.** Prepare fruits and vegetables away from raw meat and poultry and from cooked and ready-to-eat food. Clean and sanitize the work space and all utensils that will be used during preparation.

○ **Wash fruits and vegetables thoroughly under running water before cutting, cooking, or combining with other ingredients.** Pay particular attention to leafy greens, such as lettuce and spinach. Remove the outer leaves, and pull lettuce and spinach completely apart and rinse thoroughly.

○ **Refrigerate and hold cut melons at 41°F (5°C) or lower.** Discard cut melons after four hours if they have not been held at this temperature.

○ **Do not use sulfites (preservatives to maintain freshness) on any produce that will be served raw.**

Ice

○ **Ice used as food or used to chill food must be made from drinking water.**

○ **Ice that has been used to chill food should never be used as an ingredient in food or to chill beverages.**

○ **Use a clean, sanitized container and ice scoop to transfer ice from an ice machine to other containers.** Never hold or transport ice in containers that have held raw meat, fish, or poultry. Store ice scoops outside of the ice machine in a clean, protected location. Never use a glass to scoop ice, and never let your hands come in contact with the ice.

COOKING FOOD

Handling food safely prior to cooking is very important. While cooking food to the required minimum internal temperature is the only way to kill microorganisms, it does not destroy spores or toxins that microorganisms create. Safe handling of the product before it is cooked is essential to preventing microorganisms from growing and producing spores and toxins.

The minimum internal temperature at which microorganisms are destroyed varies depending upon the food. Minimum internal cooking standards have been developed for most foods and are shown in the following table. These temperatures must be reached and held for the specified amount of time. Use properly calibrated thermometers to measure the internal temperature of food. Measure internal temperatures in the thickest part of the food, and take at least two readings in different locations.

Cooking Requirements for Specific Foods		
Product	**Minimum Internal Cooking Temperature**	**Other Cooking Requirements and Recommendations**
Poultry	165°F (74°C) for 15 seconds	Poultry has more types and higher counts of microorganisms than other meats and therefore should be cooked more thoroughly.
Stuffing, Stuffed Meats, Casseroles and dishes combining raw and cooked food	165°F (74°C) for 15 seconds	Stuffing acts as an insulator, preventing heat from reaching the meat's center. Stuffing should be cooked separately.
Pork pork chops	145°F (63°C) for 15 seconds	This temperature is high enough to destroy *Trichinella* larvae that may have infested the pork.
Ground or Flaked Meats hamburger, ground pork, flaked fish, ground game animals, sausage, injected meats	155°F (68°C) for 15 seconds	Grinding meat mixes the microorganisms found on the surface throughout the meat, so thorough cooking is a must. Alternative minimum internal cooking temperatures for ground meats: 155°F (68°C) for 15 seconds 150°F (66°C) for 1 minute 145°F (63°C) for 3 minutes
Beef and Pork Roasts	145°F (63°C) for 3 minutes	Alternative minimum internal cooking temperatures for beef and pork roasts: 130°F (54°C) for 121 minutes 132°F (57°C) for 77 minutes 134°F (57°C) for 47 minutes 136°F (58°C) for 32 minutes 138°F (59°C) for 19 minutes 140°F (60°C) for 12 minutes 142°F (61°C) for 8 minutes 144°F (62°C) for 5 minutes 145°F (63°C) for 3 minutes
Beef Steaks **Veal** **Lamb** **Commercially Raised Game Animals**	145°F (63°C) for 15 seconds	

Cooking Requirements for Specific Foods		
Product	**Minimum Internal Cooking Temperature**	**Other Cooking Requirements and Recommendations**
Fish	145°F (63°C) for 15 seconds	Stuffed fish should be cooked to 165°F (74°C) for 15 seconds.
Foods Containing Fish	145°F (63°C) for 15 seconds	Fish that has been ground, chopped, or minced should be cooked to 155°F (68°C) for 15 seconds.
Shell Eggs (for immediate service)	145°F (63°C) for 15 seconds	When cooking eggs to order, only take out as many as you need. Never stack egg trays (flats) near the grill or stove.
		If eggs are cooked and held for later service, they must be cooked to 155°F (68°C) for 15 seconds and then held at 140°F (60°C).
		Discard eggs not held at or above this temperature after four hours.
		Egg dishes must be cooked to 165°F (74°C).
Vegetables		Vegetables that are cooked and held for service must be held at 140°F (60°C).
Potentially Hazardous Foods Cooked in Microwave meat, poultry, fish, eggs	165°F (74°C); let food stand for 2 minutes after cooking	Cover the food. Rotate or stir it halfway through the cooking process. Allow the food to stand covered for two minutes after cooking. Check the internal temperature in several places.

Blast chiller

Ice-water bath

Shallow pans

Reduce portion size

Acceptable Methods of
Cooling Food

COOLING FOOD

When cooked food will not be served right away, it must be cooled as quickly as possible. There are two acceptable methods of cooling food.

○ **One-stage (four hour) method:** Cool hot cooked food from 140°F to 41°F (60°C to 5°C) within four hours.

○ **Two-stage method** (Recommended by the FDA Model Food Code): Cool hot cooked food from 140°F to 70°F (60°C to 21°C) within two hours, and then to 41°F (5°C) or lower in an additional four hours for a total cooling time of six hours.

Note: The reason that the two-stage method allows six hours to cool is that in the first two hours of cooling the food is passed through the most dangerous part of the temperature danger zone where the growth of microorganisms is ideal. It is important to remember that you don't literally have six hours to cool food to 41°F (5°C). You have two hours to bring the temperature of the food to 70°F (60°C) and then four hours to bring the temperature of the food from 70°F (60°C) to 41°F (5°C) or lower.

Methods of Cooling Food

Several factors affect how fast food will cool. These include:

○ **The size of the food being cooled.** The thickness of the food or distance to its center plays the biggest part in how fast it cools. For example, a large stockpot of beef stew may take four times as long to cool as a pot half the size.

○ **How dense the food is.** The denser the food, the slower it will cool. For example, refried beans will take longer to cool than vegetable broth since the beans are denser.

○ **The container in which a food is stored.** Stainless steel transfers heat from foods faster than plastic. Shallow pans allow the heat from food to disperse faster than deep pans.

Placing hot food in a refrigerator or freezer to cool it may not move food through the temperature danger zone quickly enough. It may also raise the temperature of surrounding food items, placing them in the temperature danger zone. There are a number of methods that can be used to cool foods quickly. Any one of these methods, or a combination of them, will properly cool foods.

○ **Reduce the quantity of the food you are cooling.** Cut large food items into smaller pieces or divide large containers of food into smaller containers.

○ **Use blast chillers or tumble chillers to cool food before placing it into refrigerated storage.**

○ **Use ice-water baths.** Divide the cooked food into shallow pans or smaller pots. Place them in ice water and stir frequently.

○ **Add ice or cool water as an ingredient.** This method works for recipes that require water as an ingredient, such a soup or stew. The recipe can initially be prepared with less water than is required. Cold water or ice can then be added after cooking to cool the product and to provide the remaining water required by the recipe.

○ **Use a steam-jacketed kettle as a cooler.** Simply run cold water through the jacket to cool the food in the kettle.

○ **Stir foods to cool them faster and more evenly.** Some manufacturers make plastic paddles that can be filled with water and frozen. Stirring food with these cold paddles chills food very quickly.

Ice-Water Bath Method
Ice-water baths can be used to cool foods quickly.

STORING COOKED FOOD

If foods have not been completely cooled before storage, they can be placed in shallow pans and stored on the top shelves in the refrigerator. Leave shallow pans uncovered if they are protected from overhead contaminants. Once the food items have cooled to 41°F (5°C) or lower, they can be covered tightly. The pans of food should be positioned so air can circulate around them.

Follow FIFO when storing food. On each storage container, write the date the product was stored after preparation. Shelve food so that older food is used first. Regularly check the dates, and discard food that has exceeded its maximum storage time.

REHEATING POTENTIALLY HAZARDOUS FOOD

When previously cooked food is reheated for hot holding, take it through the temperature danger zone as quickly as possible. Food must be reheated to an internal temperature of 165°F (74°C) for fifteen seconds within two hours. If the food has not reached 165°F (74°C) for fifteen seconds within two hours, discard it.

Foods that are reheated for immediate service to a customer, such as the beef on a roast beef sandwich, may be served at any temperature, as long as the beef was properly cooked.

TAKING CORRECTIVE ACTION

Corrective actions need to be taken if time and temperature standards are not met when cooking, cooling, and reheating food.

○ **Discard food that spends more than four hours total in the temperature danger zone (41°F to 140°F [5°C to 60°C]).** This includes time spent during receiving, storage, preparation, and cooking, and again during holding, cooling, and reheating.

○ **When using the one-stage (four-hour) cooling method:** If the food has not reached 41°F (5°C) within four hours, it must be reheated to 165°F for fifteen seconds within two hours or discarded.

○ **When using the two-stage cooling method:** If the food has not reached 70°F within two hours, it must be reheated to 165°F for fifteen seconds or discarded within two hours.

○ **When reheating food:** Discard food that is being reheated if it has not reached 165°F (74°C) for fifteen seconds within two hours.

SUMMARY

The two leading causes of foodborne illness are temperature abuse and cross-contamination. Observe the four-hour rule. Food should not remain in the temperature danger zone for more than four hours total, including receiving, storage, preparation, and cooking, and again during holding, cooling, and reheating. To prevent cross-contamination, prepare raw meats, fish, and poultry in separate areas from produce or cooked and ready-to-eat foods. Use clean, sanitized utensils and work areas. Assign specific equipment to specific tasks.

Thaw frozen food in a refrigerator, under running water, in a microwave, or as part of the cooking process. When cooking, make sure that food reaches its required minimum internal cooking temperature. Before storing hot food for later use, cool it from 140°F to 70°F (60°C to 21°C) within two hours, and then to 41°F (5°C) or lower in an additional four hours. Food can be cooled by dividing it into smaller portions, using blast chillers or ice-water baths, and by stirring it with cold paddles.

Use caution when storing hot food directly in a refrigerator or freezer. Doing this may not cool the food quickly enough and may raise the temperature of the unit and the temperature of other stored items.

When reheating potentially hazardous foods that have been previously cooked, make sure the food reaches an internal temperature of 165°F (74°C) for fifteen seconds within two hours. If the food has not reached 165°F (74°C) for fifteen seconds within two hours, discard it.

NOTES

ACTIVITY

Crossword

Across:

1. The required cooking temperature that food must reach in order to kill microorganisms that can cause foodborne illness. (Abbreviation)

3. Preservatives used to maintain freshness.

8. The process of cooling hot cooked food to 41°F (5°C) or lower within four hours.

9. A plastic wand that is filled with water and frozen. When used to stir hot foods, it cools the food quickly.

Down:

2. The process of cooling hot cooked food from 140°F to 70°F (60°C to 21°C) within two hours, and then to 41°F (5°C) or lower in an additional four hours for a total cooling time of six hours.

4. The condition that exists when a potentially hazardous food is exposed to the temperature danger zone of 41°F to 140°F (5°C to 63°C).

5. A method of cooling food in which a container holding hot food is placed inside a larger container of ice water.

6. Rule that states that potentially hazardous food can spend only a fixed amount of time in the temperature danger zone before it must be discarded.

7. A device used to cool hot food quickly before it is placed in a storage cooler.

10. The process of combining several of eggs in a common container.

ACTIVITY

Word Find

Find the terms that go with the clues below.

Clues:

1. The required cooking temperature that food must reach in order to kill microorganisms that can cause foodborne illness. (Abbreviation)

2. The process of cooling hot cooked food from 140°F to 70°F (60°C to 21°C) within two hours, and then to 41°F (5°C) or lower in an additional four hours for a total cooling time of six hours.

3. Preservatives used to maintain freshness.

4. The condition that exists when a potentially hazardous food is exposed to the temperature danger zone of 41°F to 140°F (5°C to 63°C).

```
W T R O S K E S U B A E R U T A R E P M E T S M
P V L B J C W Q K R E W L W V V K K G C F U O I
B O N R X Y F M L C X W Q B O E F N Q R L N R N
L T O D C Q G G A L X I Q L E C Z Q E F E X A I
A S W L A N Z L A F A W L P X J F X I S F I M M
S U W O I Q V J R V S F U R O K G T T R W H G U
T P L N S N L Z M D N Y D B D E E A N Y B A T M
C Y X N M T G G V P T M W Y K S G T N X W X I I
H N H R E I A E X T V J W N U E P E S V V U G N
I P X D R D V G N G P I K A C S O M R X Q I T T
L G M W X O O G E D Y P P O N E X T I L F G M E
L H N V P S F O P C K D O U F E L J S V K V S R
E J Z T I J N W N G O L O P W K L V V A T C E N
R C Z C C J Y G H G I O C C P O D O M H P M L A
Y U C D E J T U K N Q J L O J Y L K H B F H U L
T U M N H Q V Y G C S C E I L G L J D D D L R T
H W Z B D S Q M R K C Z H N N D O F Z C O T R E
X V F X Z O E L A S A Z J L W G P F P U V Q U M
A T A D O T T D J S A X X X S X M A L D N Q O P
H Y N M H O F N V N T M C J Z H V E D W W T H O
T D C O P N F R X G R P S I L I B X T D T Q R V
E F D A V O J Q D I M V D P P P S P Z H L T U S
Z K L H J O L C N L R V Q X T V W R G R O E O W
I C E W A T E R B A T H C T P D C N Z T C D F W
```

5. A method of cooling food in which a container holding hot food is placed inside a larger container of ice water.

6. Rule that states that potentially hazardous food can spend only a fixed amount of time in the temperature danger zone before it must be discarded.

7. A device used to cool hot food quickly before it is placed in a storage cooler.

8. The process of cooling hot cooked food to 41°F (5°C) or lower within four hours.

9. A plastic wand that is filled with water and frozen. When used to stir hot foods, it cools the food quickly.

10. The process of combining several of this product in a common container.

ACTIVITY

What's Wrong with This Picture?

The picture below shows a situation involving food preparation. How many things can you find wrong with the picture?

A CASE IN POINT I

Case Study

On Friday, John went to work at The Fish House knowing he had a lot to do. After changing clothes and punching in, he grabbed a bus tub off the counter in the dish room and emptied it. He brought it into the refrigerator and scooped the ice from the crates of fresh fish into it. He dumped the old ice in the dish room sink and rinsed out the bus tub. Then he used the bus tub to transfer fresh ice into the ice flaker. When the bus tub was filled with flaked ice, he took it into the refrigerator and emptied it into the crates of fresh fish. When he was finished, he left the bus tub next to the ice machine.

Next, John took a case of frozen raw shrimp out of the freezer. To thaw it quickly, he put the block of frozen shrimp into the prep sink and turned on the hot water. While waiting for the shrimp to thaw, John took several fresh whole fish out of the refrigerator. He brought them back to the prep area and began to clean and fillet them. When he finished, he put the fillets in a pan in the refrigerator. He rinsed off the boning knife and cutting board in the sink and wiped off the worktable with a dish towel.

Next, John transferred the shrimp from the sink to the worktable in a large plastic bucket. He peeled, deveined, and butterflied the shrimp with the boning knife. He put the prepared shrimp in a covered container in the refrigerator, then started preparing fresh produce.

What did John do wrong?

A CASE IN POINT II

Case Study

By 7:30 in the evening, all the residents at Sunnydale nursing home had eaten dinner. As she began cleaning up, Angie realized that she had a lot of chicken breasts left over (see *A Case in Point I* in Chapter 6). Betty, the new assistant manager, had forgotten to inform Angie that several residents were going to a local festival and would miss dinner.

"No problem," Angie thought. "We can use the leftover chicken to make chicken salad."

Angie left the chicken breasts in a pan on the steam table while she started putting other foods away and cleaning up the kitchen. At 9:45 p.m. when everything else was clean, she put her hand over the pan of chicken breasts and decided they were cool enough to handle. She took the pan out of the steam table, covered it with plastic wrap, and put it in the refrigerator.

Three days later, when she came in to work on the early shift, Angie decided to make chicken salad from the leftover chicken breasts. After she hung up her coat and put on her apron, Angie took all the ingredients she needed for chicken salad out of the refrigerator and put them on a worktable. Then she started breakfast.

First, she cracked three dozen eggs into a large bowl, added some milk, and set the bowl near the stove. Then she took a slab of bacon out of the refrigerator and put it on the worktable next to the chicken salad ingredients. She peeled off strips of bacon onto a sheet pan and put the pan into the oven. After wiping her hands on her apron, she went back to the stove to whisk the eggs and pour them onto the griddle. When they were almost done, Angie scooped the scrambled eggs into a hotel pan and put it in the steam table.

As soon as breakfast was cooked, Angie went back to the worktable to wash and cut up celery and cut up the chicken for chicken salad.

What did Angie do wrong?

NOTES

TRAINING TIPS

Training Tips for the Classroom

1. Group Breakout Activity: SOP for Menu Items

Objective: *After completing this activity, class participants will be able to develop standard operating procedures (SOPs) for preparing food and be able to identify some Critical Control Points for various food groups. Note: this higher level activity may not be suitable for all classes.*

Directions: Make a list of six to eight different menu items, including at least one item from each major food category discussed in Chapter 7 under "Preparing Food." A representative list might include:

○ Fruits and Vegetables: Chef's Salad

○ Meats, Fish, and Poultry: Fresh Whitefish, Roast Turkey, Loin of Pork

○ Protein Salads and Sandwiches: Chicken Salad

○ Eggs: Denver Omelet

○ Batters and Breading: Hand-Breaded Fried Chicken

Break the class into groups of about three people each. Give each group a sheet of paper with one of these menu items printed at the top. Each group will work on a different menu item.

Ask each group to create a recipe for its assigned menu item. The recipe should include ingredients, preparation steps, and, especially, sanitation instructions. (Note: Accurate weights and measures are not necessary for these recipes, since this activity should focus on food-safety guidelines rather than culinary accuracy.)

Give the groups a time limit to develop recipes for their items. Remind them that these recipes should focus on Critical Control Points and SOPs for safe food handling.

When time is up, have each group present its recipe. Allow time for discussion from the class. Make necessary changes and additions. Distribute copies of the recipes as a handout at the end of the day.

2. "Chill Out!" Team Contest

Objective: *After completing this activity, class participants will be able to identify the various methods of cooling food, discuss advantages and disadvantages, and identify foods best suited for each cooling method.*

Note: Before this activity, remind students that failure to cool foods properly is the number-one cause of outbreaks of foodborne illness. Present this exercise after reviewing the critical limits for cooling foods, but before a discussion of cooling methods.

Directions: Print up a form that asks students to list as many cooling methods as they can think of, along with a list of food types best suited for each method, and advantages and disadvantages of each method.

Divide the class into teams of two and give each team one form. Allow about seven minutes (closed book) for the teams to list cooling methods, food types, and advantages and disadvantages.

After time is up, let the teams grade their sheets as follows:

- ○ For each cooling method (recognized as safe)   5 points
- ○ For each type of food to be cooled by this method   2 points
- ○ Each advantage of the cooling method   3 points
- ○ Each disadvantage of the cooling method   3 points

The team with the most point wins.

Allow time for discussion, and let students make additions and correct errors on their papers. Create a master list of cooling methods from this activity as a reference handout.

3. "Cook-Kill" Temperature Quiz

Objective: *After completing this activity, class participants will be able to identify safe internal cooking temperatures for different food groups.*

Directions: As a pop quiz, give students a printed list of twelve to fifteen varied menu items, with a blank space next to each item. Ask students to fill in the minimum safe internal temperature for cooking each item. Put in some trick menu items, such as rare roast beef, stuffed ostrich cutlets, buffalo burger.

Limit this activity to two minutes. Review the different temperatures with the class. See if anyone in the class got a perfect score.

Use this quiz as a springboard to a group discussion on safe cooking temperatures for different food groups. Discuss how to monitor cooking temperatures (visual check versus thermometers or timers) and how often such monitoring is necessary. Discuss how one might serve food that does not reach these critical limits, due to customer requests (rare hamburger) or type of item (sushi, raw oysters).

Training Tips on the Job

1. "Cool It" Cooling Activity

Purpose: *To examine the methods used in your operation to cool food and to identify any problems with your current practices. When your kitchen staff is involved in this analysis, they can be the ones to identify any problems and will buy into the decision to change methods if necessary.*

Directions: Invite the kitchen staff to participate in a competition.

Select a regular menu item that your establishment prepares in advance and then cools down for future use. Chili is a good example. The next time you're scheduled to make chili, have the cook take the temperature of the chili just before taking it off the stove. Record this temperature, then tell the cook to cool the chili as she normally does.

Ask each staff member to guess what the temperature of the chili will be two hours after removing it from the stove, and again four hours later. Record their guesses on index cards.

Have someone take and record the temperature of the chili every hour. After the first two hours, and after six hours, verify the final temperature yourself, and then check the index cards. List the different temperatures that people guessed, and announce a winner. (If the chili does not reach 70°F (21°C) within two hours and 41°F (5°C) in less than six hours, ask if they want to eat the chili for lunch!)

Discuss with your staff the importance of cooling foods. Solicit their input to develop a system for cooling chili and other food products to 41°F (5°C) within two plus four hours. Discuss and brainstorm different methods, and then implement the best method as your standard cooling process.

2. "Cooking Pays" Temperature Quiz

Purpose: *To reinforce the importance of monitoring temperature when handling food in your operation.*

Directions: Every day, carry a number of tokens in your pocket, along with your calibrated thermometer or thermocouple. Whenever you walk through the kitchen and see an employee using a thermometer to check the temperature of the food, give him or her a token. If employees can tell you the temperature of the food product and are accurate within 2°F (1°C), give them two more tokens. During your regular staff meetings, give out prizes based on the number of tokens collected. This activity verifies that employees are using thermometers.

As you conduct this activity, remind the cooks of the importance of monitoring both time and temperature when handling potentially hazardous food, and solicit their cooperation in minimizing the time these foods are in the temperature danger zone.

3. Work Station Safety Checklists

Purpose: *To involve kitchen staff members in the creation of food-safety checklists for specific work areas in your establishment.*

Directions: Create a food-safety team from a group of volunteers from the kitchen staff. Ideally all different shifts and work areas should be represented.

Assign members of the team to study different work areas in your operation (cold prep, grill, stove, oven, bakery, cleanup, storage, and so on) and ask them to develop food-safety checklists for those specific work areas. They should keep in mind the three major food-safety concerns: monitoring time and temperature, personal hygiene, and cross-contamination.

After receiving the checklists, edit them to make them consistent with one another and easy to read. Suggest ways to make them colorful or color-coded, and translate them into several languages if necessary. Review your work with the team. Have the checklists professionally printed and laminated, and post them in the appropriate work areas for easy reference.

MULTIPLE-CHOICE STUDY QUESTIONS

1. You have only one ceramic cutting board available for food preparation. You have just sliced some chicken breasts for cooking and now need to prepare a green salad. What should you do to the cutting board before you use it for preparing the salad?

   A. Scrub it using hot, potable water and a detergent, then sanitize.
   B. Dry it with a paper towel.
   C. Rinse it under very hot water.
   D. Turn it over and use the reverse side.

2. Which of the following is not a safe method for thawing a frozen brisket of beef?

   A. Let it sit at room temperature for five hours.
   B. Put it in a microwave set on automatic defrost.
   C. Immerse it in room-temperature running water for one hour.
   D. Let it sit in the refrigerator overnight.

3. With a probe thermometer, you measure the temperature of the breast meat and the stuffing of a stuffed chicken. What would tell you that the chicken has been safely cooked?

   A. Both temperatures read 155°F (68°C).
   B. The stuffing temperature reads 160°F (71°C).
   C. Both temperatures read 165°F (74°C).
   D. The meat temperature reads 170°F (77°C).

4. You are making omelets to order for the Sunday brunch at the hotel where you work. How should you handle the eggs to ensure that you are serving safe omelets?

   A. Keep all of the eggs in their shells. Take out only the number of eggs you expect to use and store them away from the stove.
   B. Keep all of the eggs in their shells next to your cooking station.
   C. Reserve one special bowl for cracking the eggs into before cooking.
   D. Crack all of the eggs at once into a large container.

5. You are making a mixed vegetable tray for the salad bar. Which of the following is a proper procedure for preparing the vegetables to ensure food safety?

A. The tomatoes should be scrubbed with a bristle vegetable brush.
B. Each leaf of the romaine lettuce should be washed separately.
C. The carrots should be soaked in ice water.
D. The mushrooms should be dry brushed to remove any field soil.

6. Your ice maker produces ice for chilling the plates for the salad bar and for beverage service. Which is a proper procedure for dispensing the ice from the ice maker storage bin?

A. Ice for both uses should be dispensed with a handled scoop stored properly outside of the ice bin.
B. Ice for beverages must be taken from the back of the salad bar.
C. Ice for the salad bar should be dispensed in large plastic containers using a clean sanitized glass.
D. Ice for both uses should be made with carbon-filtered water.

7. You need to cool a large stockpot of clam chowder for use the following day. What is the first thing you should do?

A. Put the large pot containing soup in the refrigerator.
B. Transfer the soup to a different large pot.
C. Transfer the soup from the large pot to a shallow stainless-steel pan.
D. Put some ice cubes into the soup to help the cooling.

8. Your soup and salad bar is serving minestrone soup that was prepared the day before and refrigerated overnight. The serving table has a temperature-controlled heater that can maintain an equipment temperature of 165°F (74°C). What do you need to do to the soup to prepare it for serving?

A. Reheat the cold soup on the serving table and stir it to speed the reheating.
B. Reheat only small amounts of the soup at a time on the serving table.
C. Microwave the soup to 75°F (24°C) and then put it on the serving table to finish reheating to 165°F (74°C).
D. Reheat the soup quickly to 165°F (74°C) for fifteen seconds on the stove before putting it on the serving table.

9. Which of the following foods has been safely cooked?

   A. A rare beef roast that has been cooked to an internal temperature of 135°F (57°C)

   B. A pork roast that has been cooked to an internal temperature of 135°F (57°C)

   C. A whole turkey that has been cooked to an internal temperature of 155°F (68°C)

   D. A tuna casserole that has been cooked to an internal temperature of 165°F (74°C)

10. How should a batch of beef stew that has been cooled to 70°F (21°C) be stored for service the following day?

    A. In a shallow, loosely covered pan on the top shelf of the refrigerator

    B. In a plastic jar at the back of the refrigerator

    C. In a pot in the storage freezer

    D. In a warming oven set to 140°F (60°C)

Section 8

Protecting Food During Service

Knowledge

TEST YOUR FOOD-SAFETY KNOWLEDGE

1. **True or False:** A hot-holding device can be used for reheating foods, provided it is capable of reaching a temperature of 140°F (60°C). *(See Holding Hot Foods—Do's and Don'ts, page 8-3.)*

2. **True or False:** Raw chicken can be served next to ready-to-eat foods on a self-service buffet if care is taken during setup. *(See Self-Service Areas, page 8-8.)*

3. **True or False:** A sneeze guard protects the foods on a salad bar from cross-contamination. *(See Self-Service Areas, page 8-8.)*

4. **True or False:** Catered foods should be delivered in insulated food containers. *(See Delivery, page 8-9.)*

5. **True or False:** Foods that have been prepared safely and cooked properly in the kitchen may not necessarily be safe to eat when they reach the table. *(See Serving Foods Safely, page 8-4.)*

CONCEPTS

○ **Hot-holding equipment:** Equipment such as *bains maries,* chafing dishes, steam tables, and heated cabinets designed to hold foods at temperatures of 140°F (60°C) or above. This equipment is *not* designed to heat foods.

○ **Cold-holding equipment:** Equipment specifically designed to keep cold foods at temperatures of 41°F (5°C) or below. Includes refrigerated food bars, iced displays, refrigerated sandwich rails, and insulated carriers.

○ **Corrective action:** Action taken to correct an error or lapse in safe food handling or HACCP procedures. For example, when the temperature of a hot food falls below 140°F (60°C), the proper corrective action is to reheat the food to 165°F (74°C) for fifteen seconds within two hours.

○ **Personal hygiene:** Sanitary health habits that include keeping body, hair, and teeth clean; wearing clean clothes; and washing hands regularly, especially when handling food and beverages.

○ **Food bar:** Self-serve display or buffet table at which customers can select a variety of foods. Food bars may feature hot and cold foods.

○ **Sneeze guard:** A food shield placed in a direct line between food on display and the mouth/nose of a person of average height, usually fourteen to forty-eight inches above the food.

○ **Off-site service:** Any time food is served someplace other than where it is prepared or cooked. Off-site service includes catering, satellite feeding operations, meal delivery, and vending.

○ **Vending machine:** A machine that dispenses food. Vending machines may accept coins, currency, or debit cards and can dispense both hot and cold foods, beverages, and snacks.

○ **Mobile unit:** A portable foodservice facility. Mobile units range from simple vending carts that hold and display prepackaged foods to full field kitchens capable of preparing and cooking elaborate meals.

○ **Temporary unit:** A foodservice facility that usually operates in one location for less than fourteen days in conjunction with a special event or celebration.

○ **Single-use items:** Disposable tableware or packaged foods designed to be used only once. Single-service items include plastic flatware, paper or plastic cups, plates and bowls, as well as single-serve foods and beverages.

GENERAL RULES FOR HOLDING FOODS

○ Any conflict between food quality and food safety must always be decided in favor of food safety.

○ When holding foods for service, keep cold foods cold and hot foods hot.

○ Prepare and cook only as much food as you will use or serve in a short period of time.

Holding Hot Foods—Do's and Don'ts

Do's

○ **Use only hot-holding equipment that can keep foods at 140°F (60°C) or higher.**

○ **Stir foods at regular intervals.** Stirring will distribute heat evenly throughout the food.

○ **Keep food covered.** Covers retain heat and keep contaminants from falling into the food.

○ **Measure internal food temperatures at least every two hours.** Use a probe thermometer. Record temperatures in a log.

○ **Discard hot foods after four hours if they have not been held at or above 140°F (60°C).**

Don'ts

○ **Never use hot-holding equipment to reheat foods.** Reheat foods first to 165°F (74°C), then transfer to holding equipment.

○ **Never mix freshly prepared food with foods being held for service.** This will prevent cross-contamination.

Courtesy of Cooper Instruments Corporation, Middlefield, CT

Hot-Holding Equipment
Use only hot-holding equipment that can keep foods at 140°F (60°C) or higher.

Cold-Holding Equipment
Cold-holding equipment must keep foods at 41ºF (5ºC) or lower.

Properly Stored Utensils
If stored in food, utensils should be stored with the handle extended above the rim of the container.

Holding Cold Foods— Do's and Don'ts

Do's

○ **Use only cold-holding equipment that can keep foods at 41°F (5°C) or lower.**

○ **Measure internal food temperatures at least every two hours.** If above 41°F (5°C), take corrective action. Record temperatures (and corrective actions) in a log.

○ **Protect food from contaminants with covers or food shields.**

Don'ts

○ **Do not store foods directly on ice.** Place foods in pans or on plates first. Whole fruits and vegetables, raw cut vegetables, and molluscan shellfish are the only exceptions. Ice used on a display should be self-draining. Wash and sanitize drip pans after each use.

SERVING FOODS SAFELY

Food servers need to be just as careful as kitchen staff. Servers can contaminate food simply by handling the food-contact area of a plate. Train your kitchen staff to follow these procedures to serve food safely.

Serving Food Safely for Kitchen Staff and Servers— Do's and Don'ts

Do's

○ **Store utensils properly.** Serving utensils can be stored in the food, with the handle extended above the rim of the container. They can also be placed on a clean, sanitized food-contact surface. Spoons or scoops used to serve foods such as ice cream or mashed potatoes may be stored under running water.

○ **Use serving utensils with long handles.** A long-handled ladle will keep a server's hands away from soup in a kettle; a cup will not.

○ **Clean and sanitize utensils before using.** Use separate utensils for each food item, and properly clean and sanitize them after each task. Utensils should be cleaned and sanitized at least once every four hours during continuous use.

○ **Practice good personal hygiene.**

○ **Handle glassware and dishes properly.**

○ **Hold flatware and utensils by the handles.** Store flatware so servers will grasp the handles, not the food-contact surfaces.

○ **Serve milk from refrigerated bulk dispensers or in single-serve cartons.** Serve cream and half-and-half in single-serve containers or in covered pitchers.

○ **Use plastic or metal scoops or tongs to get ice.**

○ **Cloths used for cleaning food spills shouldn't be used for anything else.** When cleaning tables between guest seatings, wipe up spills with a disposable dry cloth. Then clean the table with a moist cloth that has been stored in a fresh detergent solution.

○ **Servers must wash their hands before handling place settings or food.**

Don'ts

○ **Never touch cooked or ready-to-eat foods with bare hands.** Always use gloves or utensils to handle these foods.

○ **Never stack glassware or dishes when serving.**

○ **Never touch food with bare hands.** Serve with tongs or gloves, for example.

○ **Never assign employees to more than one job during a shift.** Serving food, setting tables, and busing dirty dishes are separate tasks with different responsibilities.

Refer to page 8-6 for the proper techniques to carry utensils and serve food.

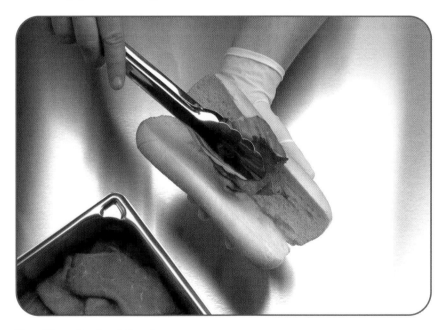

Handling Cooked Food
Never touch cooked or ready-to-eat foods with bare hands.

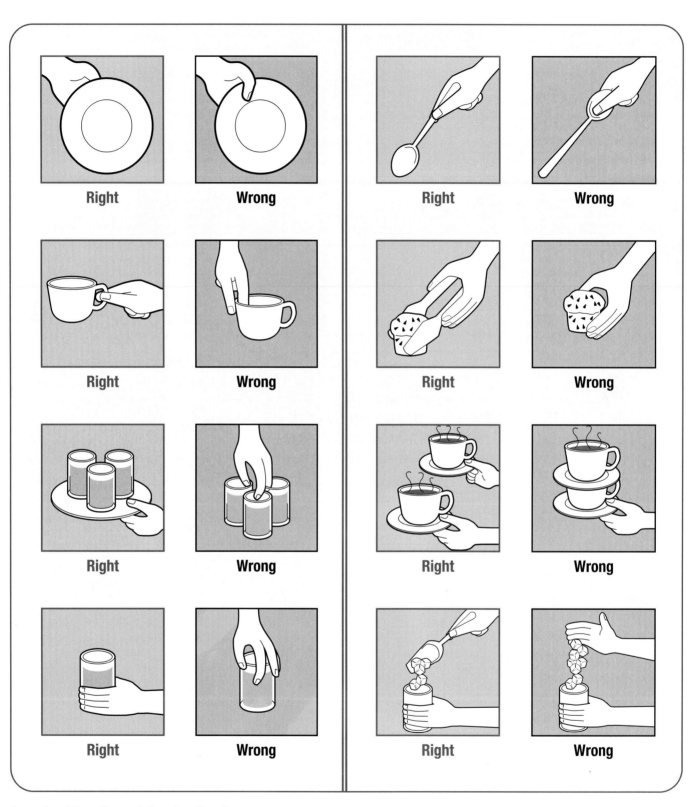

Carrying Utensils and Serving Food
Proper and improper ways to carry utensils and serve food.

Re-Serving Food

Servers and kitchen staff should know rules about re-serving foods that have been previously served to a customer.

○ **Re-serve only unopened, prepackaged foods such as condiment packets, wrapped crackers or breadsticks, and other sealed foods.** Do not re-serve anything to people at risk such as nursing home residents or hospital patients.

○ **Never re-serve plate garnishes such as fruit or pickles to another customer.** Served but unused garnish must be discarded. Never re-serve uncovered condiments.

○ **Do not re-serve bread baskets, even if they have not been used.**

○ **Don't combine leftovers with fresh food.** For example, opened portions of salsa, mayonnaise, mustard, or butter should be thrown away.

○ **Linens used to line bread baskets must be changed each time a customer is served.**

Refilling Food in Holding Equipment
Never mix freshly prepared food with foods being held for service.

Self-Service Areas

Buffets and food bars give patrons the chance to select what they want to eat. These self-service areas also can be easily contaminated. Buffets and food bars should be monitored closely by employees trained in food-safety procedures. Assign a staff member to replenish food-bar items and to hand out fresh plates for return visits. Post signs with polite tips about food-bar etiquette. Here are more rules for food bars:

Protecting Food on Display
Sneeze guards must be between fourteen and forty-eight inches above the food, in a direct line between the food and the mouth or nose of an average customer.

○ **Protect food on display with sneeze guards or food shields.** These must be between fourteen and forty-eight inches above the food, in a direct line between the food and the mouth or nose of a customer of average height.

○ **Identify all food items.** Label containers on the food bar (don't use probe-type tags in the food itself). Write the names of salad dressings on ladle handles.

○ **Keep hot food hot—140°F (60°C) and cold food cold—41°F (5°C).** Measure the internal temperature of food using a thermometer, and record temperatures at least every two hours.

○ **Replenish foods on a timely basis.** Practice the FIFO method of product rotation. Prepare and replenish small amounts at a time so food is fresher and has less chance of being exposed to contamination. Never mix fresh food with food being replaced.

○ **Keep raw foods separate from cooked and ready-to-eat foods.** Customers can easily spill foods when serving themselves. Use separate displays or food bars for raw foods and cooked foods so there's less chance of cross-contamination.

○ **Do not let customers use soiled plates or silverware for refills.** Customers can use glassware for refills as long as beverage dispensing equipment doesn't come in contact with the rim or interior of the glass.

Sneeze guards placed at a proper height between the food and the mouth and nose of a customer of average height can help protect food, but can't prevent all forms of cross-contamination.

OFF-SITE SERVICE

Delivery

Many establishments such as schools, hospitals, caterers, and even restaurants prepare food at one location and then deliver it to remote locations. The greater the time and distance from the point of preparation to the point of consumption, the greater the risk that food will be exposed to contamination or temperature abuse.

○ **Use rigid, insulated food containers capable of maintaining food temperatures above 140°F (60°C) or below 41°F (5°C).** Containers should be sectioned so foods don't mix, leak, or spill. They must also allow for air circulation to keep temperatures even. Keep containers clean and sanitized.

○ **Clean the inside of delivery vehicles regularly.**

○ **Practice good personal hygiene when distributing food.**

○ **Check internal food temperatures regularly.** Take corrective action if food isn't at the proper temperature. Within two hours, reheat hot foods to 165°F (74°C) for fifteen seconds before serving. Re-chill cold foods to 41°F (5°C) or below within four hours. If containers or delivery vehicles aren't maintaining proper food temperatures at the end of each route, reevaluate the length of delivery routes or the efficiency of the equipment you use.

○ **Label foods with instructions for proper storage, shelf life, and reheating for employees at off-site locations.**

○ **Provide food-safety guidelines for consumers.** If you or your employees won't be serving the foods you deliver, provide customers with information on which items should be eaten immediately, which may be saved for later, and how to serve them.

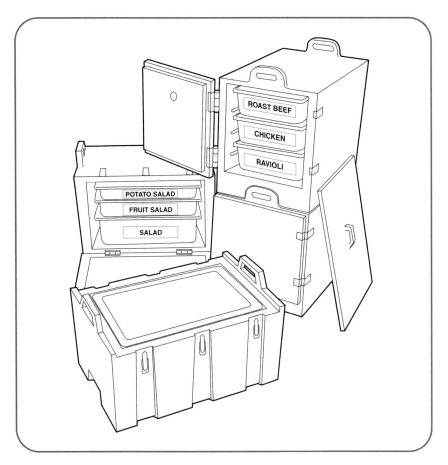

ROAST BEEF
CHICKEN
RAVIOLI
POTATO SALAD
FRUIT SALAD
SALAD

Delivery of Food Off-Site
Equipment used to transport food must be designed to maintain safe food temperatures and must be easy to keep clean.

Catering

Caterers provide food for private parties and events, and public and corporate functions. They may bring in prepared food, or cook food on-site in a mobile unit, a temporary unit, or in the customer's own kitchen.

○ Make sure there is safe drinking water for cooking, warewashing, and handwashing, as well as adequate power for holding and cooking equipment.

○ Deliver raw meats frozen and wrapped, on ice. Deliver milk and dairy products in a refrigerated vehicle or on ice. Use insulated containers for all potentially hazardous foods. Keep them cold with ice or frozen gel-ice packs.

○ Serve cold foods in containers on ice. If that isn't possible, record the time when the potentially hazardous food was first taken out of cold storage, and then discard the food after four hours.

○ Keep raw and ready-to-eat products separate. For example, store raw chicken separately from ready-to-eat salads.

○ Use only single-use items. Make sure customers get a new set of disposable tableware for refills. Arrange proper garbage disposal away from food-prep and serving areas.

○ Provide instructions for proper storage, shelf life, and reheating if food is left with the customer after the event.

Vending Machines

Food prepared and packaged for vending machines has to be handled with the same care as any other food served to a customer. Vending operators also have to protect vended foods from contamination and temperature abuse during transport, delivery, and service.

○ **Keep foods at the right temperature**—41°F (5°C) or below and 140°F (60°C) or above.

○ **Machines must have automatic cutoff controls** that prevent foods from being dispensed if the temperature stays in the danger zone for a certain amount of time.

○ **Check product shelf life daily.** Replace foods with expired code dates. If refrigerated foods aren't used within seven days after preparation, they must be thrown out.

○ **Dispense potentially hazardous foods in their original containers.** Fresh fruits with an edible peel should be washed and wrapped before being put in a machine.

EIGHT RULES OF SAFE FOOD HANDLING

All employees should know the eight basic rules of safe food handling, and should be responsible for safe food practices in their assigned areas.

1. **Practice strict personal hygiene.** Employees should wash their hands regularly and never touch ready-to-eat food with their bare hands. Employees who are ill should not report to work.

2. **Monitor time and temperature and prevent cross-contamination** when storing and handling food during preparation.

3. **Make sure raw products are kept separate from ready-to-eat foods.**

4. **Avoid cross-contamination by cleaning and sanitizing food-contact surfaces,** equipment, and utensils before and after every use, and at least once every four hours during continuous use.

5. **Cook foods to their required minimum internal temperature or higher.**

6. **Hold hot foods at 140°F (60°C) or above and cold foods at 41°F (5°C) or below.**

7. **Chill cooked food to 41°F (5°C) within four hours.** (Alternatively, cool cooked food to 70°F (21°C) within two hours, and then chill to 41°F (5°C) within four hours.)

8. **Reheat foods for service to an internal temperature of 165°F (74°C) for fifteen seconds within two hours.**

SUMMARY

Safe foodhandling practices don't stop once food is properly prepared and cooked. To make sure the food you serve is safe, you must continue to protect it from temperature abuse and contamination until it is eaten.

When holding foods for service, keep hot foods hot and cold foods cold. Make sure all employees, including servers, practice good personal hygiene. Train employees to avoid cross-contamination when handling utensils, service items, and tableware. Customers, too, can unknowingly contaminate foods in self-service areas such as food bars and buffets. Take special precautions when preparing, delivering, or serving food off-site. Finally, make sure all employees know and understand the eight rules of safe food handling.

NOTES

ACTIVITY

Crossword Puzzle

Across:

5. A buffet or display table where customers can serve themselves from a variety of hot or cold foods.

6. A service that dispenses food and beverages from coin-operated machines.

7. Types of equipment used to maintain food temperatures of 140°F (60°C) or above.

8. A service that provides food for parties, events, and special functions.

9. Flatware, dinnerware, or food packaged in containers designed to be used once and discarded.

10. Restaurant and foodservice facilities that are set up and operated for short periods of time at special events such as fairs and sports tournaments.

Down:

1. Steps you should take when food doesn't meet standards at a Critical Control Point (CCP).

2. Portable facilities for preparing, cooking, or serving food.

3. A shield placed between food and customers' faces on self-service displays.

4. A state of cleanliness and appearance that helps prevent cross-contamination.

ACTIVITY

Word Find

Find the terms that go with the clues below.

Clues

1. Types of equipment used to maintain food temperatures of 140°F (60°C) or above.

2. Steps you should take when food doesn't meet standards at a Critical Control Point (CCP).

3. A state of cleanliness and appearance that helps prevent cross-contamination.

4. A buffet or display table where customers can serve themselves from a variety of hot or cold foods.

5. A shield placed between food and customers' faces on self-service displays.

6. A service that provides food for parties, events, and special functions.

```
P E R S O N A L H Y G I E N E S C Z A K N Y K N
N R L V I X L A J S M E T I E S U E L G N I S T
Y G M A Z I U E Z F H V C W S D R A K Q T Y I F
D N P F X M B K B U O T O H Q U K I E X T N L R
H R M W J X Q R M C T O X X D Z C C P I U I O X
Z O A W M Z R L Q Z K K Y O W O X H N Y N S Y U
A O T U G N W S J L Y F R B E S O U R O V V Y S
E M X H G W P Y Z U P D E Y R Z E A I E Q B G F
T U S O O E B X B D J U P Q N L R T V P S J R F
F H N Q S L Z P M M T M W O I O C T H J H M F N
W Z Z E U D D E M K E K F B P A B P B Q U Y X Q
W Y J L W M B I E H W X O M E Y S R R F O W F Y
A T A I Z O V I N N J M E V D U U V M H S F I Y
C C F S R T D O H G S T I H L U B D C W R Z T V
V D C O X Q L U X T E T H U B M S Z V N Y M J R
C O V C G N Q A Q R C Q T N P P S M D U N G V O
D Y G B K Z F T X E N P U G L E B G A J U Y S L
S C A N R Y W R R V S C Z I N G E T Z T U F Q K
C X A U O W A R X R J K C Y P I N U M V H O U S
X I G O Q B O Z O H Q Z O J Z M R I U V T A S U
V F B F D C O J P G B G G U E C E E D B C P B D
C U P O I E Y Z U F B Z L N C W C N T N U H Z F
Q K O C I R Y G L K G Q F X O B P E T A E X C T
A F B A D S O D R H S Z Y C R H M S U H C V U Z
```

7. A service that dispenses food and beverages from coin-operated machines.

8. Portable facilities for preparing, cooking or serving food.

9. Restaurant and foodservice facilities that are set up and operated for short periods of time at special events such as fairs and sports tournaments.

10. Flatware, dinnerware, or food packaged in containers designed to be used once and discarded.

ACTIVITY

New Employee Training

This activity can be conducted in pairs working independently or as a role-playing exercise in front of a group. Each participant's approach to the problem should be critiqued at the end of the activity.

Directions: Ask participants to read the case study below. Assign the roles of **"Megan"** and **"Manager"** and give them the following guidelines:

Megan: Analyze your behavior and explain why you did what you did. Tell the manager if you think you did anything wrong.

Manager: Determine what Megan did wrong. Ask if she learned any behavior from watching experienced employees. Then walk Megan through the proper procedures and develop a training module for all servers.

A CASE IN POINT

Case Study

Megan, a new server at The Fish House, reported for work ten minutes early on Thursday. Excited about her new job, she had made sure to shower and wash her hair before going to work. When she arrived, she changed into a clean uniform, pulled her hair back tightly into a ponytail, and checked her appearance. She had used makeup sparingly and wore only small hoop earrings and a short chain necklace for jewelry.

Her shift started well. One of her customers ordered a menu item Megan hadn't tried yet. When the order came up, Megan dipped her finger into the sauce at the edge of the plate for a taste. As the shift progressed, Megan's station got busier. When the hostess came to ask how soon one of Megan's tables could be cleared, Megan decided to do it herself. She took the dirty dishes to a bus station, then wiped down the table with a serving cloth she kept in her apron.

The buser, John, finished another task and came to help her set the table. While he put out silverware and linens, Megan filled water glasses with ice by scooping the glasses into the ice bin in the bus station. Only one slice of bread and one pat of butter were missing from the basket that had been on the table, so she put that on a tray with the water glasses and took it to the table.

The table was reset in record time, and Megan soon had more guests. While taking their orders, Megan reached up to scratch a mosquito bite on her neck. Then she went to the kitchen to turn in the order and pick up a dessert order for another table.

ACTIVITY

Serve or Discard?

Indicate whether the foods listed below can be re-served (R) or must be discarded (D).

_____   Untouched fresh bread basket

_____   Individually wrapped crackers

_____   Butter plate

_____   Mustard packet

_____   Bowl of mayonnaise

_____   Unopened single-serve yogurt cup in nursing home

_____   Ice used on food bar

_____   Chilled cooked food reheated to 165°F (74°C) for fifteen seconds

_____   Covered cream container left on table for the day

_____   Fruit garnish

ACTIVITY

Right or Wrong?

The pictures below show ways food and utensils might be served and handled. Below each picture, indicate whether the procedure is right or wrong and explain why.

1.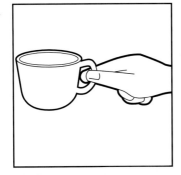

2.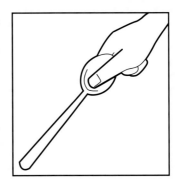

3.

4.

5.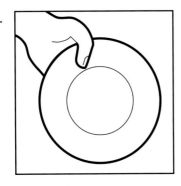

6.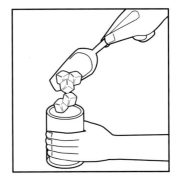

ACTIVITY

What Service Mistake Caused the Illness?

Scenario: Fifty people from a local car dealership celebrated their Christmas party at the Blue Dragon, a Mongolian barbecue and buffet. Approximately one hour after eating, about forty people became ill, suffering from severe vomiting. You now have the opportunity to ask the health inspector (your instructor) specific questions related to the incident. Use the following facts to determine the mistake made during service.

○ The buffet was stocked with both raw and ready-to-eat food.

○ The buffet attendant was ill.

○ The chicken fried rice was a popular item that night and had to be restocked often.

○ The restaurant is well known for its shrimp dish, which is the spiciest in town.

○ The buffet attendant took the temperature of foods on the buffet every two hours.

○ The dishwasher called in sick that evening, further burdening an already short staff.

TRAINING TIPS

Training Tips for the Classroom

1. Group Activity: Food Contamination During Service

Objective: *After completing this activity, class participants will be able to identify ways food can be contaminated during service; standard operating procedures (SOPs) and Critical Control Points (CCPs) that can minimize contamination; and corrective actions that should be taken for various modes of food service.*

Directions: Break the group into teams, and assign one team to each of the following topics about safe foodhandling that were discussed in Section 8.

- hot holding
- cold holding
- table service and re-service of food
- food bars, self-service, buffets
- off-site delivery
- carryout
- catering
- mobile units
- temporary units
- vending machines

Have the teams address the following questions about the assigned topics.

- How can food become contaminated at this stage or under these conditions?
- What standard operating procedures (SOPs) must be in place to minimize food contamination?
- When you look at the flow of a particular food, what is the last point where a hazard can be eliminated, prevented, or reduced?
- How do you verify that food-safety procedures have been followed and standards have been met?
- What corrective actions and documentation must be performed, and who will perform them?

Allow about ten minutes for the teams to answer the questions, and allow time for each team to present its answers to the class. Then ask the entire class the following questions, to answer in group discussion.

- Which team's information applies to your establishment?
- Does your establishment have these systems and procedures in place?
- If not, how might they be put in place?

2. The Eight Rules of Safe Food Handling: A Self-Assessment Test

Objective: *After completing this activity, class participants will be able to assess foodhandling practices in their establishments, discover any weaknesses, and identify systems and procedures that could be put in place to improve foodhandling practices.*

Directions: Ask students to honestly assess how well their establishments follow the Eight Rules of Safe Food Handling listed at the end of Section 8.

Print a rating sheet with a list of the Eight Rules of Safe Food Handling. Next to each rule, create a rating scale from one to four. Ask students to rate foodhandling practices at their own establishments on a scale of one to four, one (1) being "perfect" and four (4) being "poor" or "needs work." The sheet should look like this.

Perfect	Good	Average	Poor	
1	2	3	4	
—	×	—	—	Keep raw products separate from ready-to-eat foods.
—	—	×	—	Chill cooked food to 41°F (5°C) within four hours.

Allow ten or fifteen minutes for this self-assessment test, then solicit responses from the class for each of the rules. Ask for a show of hands: how many students reported a one, two, three, or four on each safe foodhandling rule? Record the results on the worksheet, and report the results orally, or write the tally on a blackboard or flipchart if one is available.

Then allow time for in-depth discussion of each rule. If any students indicated a score of "average" or "poor" on one or more of the rules, ask them the following questions.

○ What food-safety system and procedures do you currently have in place regarding the foodhandling rule?

○ What additional food-safety systems and procedures could you put in place to improve food handling?

Solicit class input to help improve foodhandling practices at the establishment in question.

For students who scored "good" or "perfect" on any rule, ask if their establishment ever falls short of the rule, and ask if there is anything they could do to further improve current foodhandling practices.

Some students may be reluctant to report anything negative about their establishment, so you should ask for volunteers to answer some of these questions. The idea is to have a lively class discussion about each of the rules, not to make anyone uncomfortable.

Remind the class that while they may find their establishments to be safe in general, there is always room for improvement. We should take every opportunity to improve systems and procedures to minimize food contamination.

Training Tips on the Job

1. *Front-of-the-House Food-Safety Task Force*

Objective: *To involve service staff in an assessment of foodhandling practices in the front-of-the-house and to solicit recommendations for improving current practices.*

Directions: Develop a food-safety task force made up of the front-of-the-house personnel. Include at least one person from every front-of-the-house position in the task force. This should include servers, busers, cashiers, bartenders, hosts and hostesses, catering personnel, delivery drivers, shift supervisors, and managers.

Plan to hold three or more meetings with your food-safety task force. Before your first meeting, do some research. Get information from state and local health departments about outbreaks of foodborne illness that have occurred in recent years. Ask which were directly caused by breakdowns in temperature control, personal hygiene, or cross-contamination, or due to mistakes during holding, service, or transportation.

First Meeting: Start your first meeting by presenting the information about recent outbreaks of foodborne illness. Then ask members of your food-safety team to look for and document hazards and food-safety violations in their particular areas of the establishment. Schedule another meeting to hear their reports.

Second Meeting: Your next meeting should allow time for each person to report on the hazards or safety violations found in his or her work area. Encourage and allow time for group feedback. The next assignment for the task force is to research possible solutions to problem areas. Suggestions for possible solutions can come from fellow employees and other foodservice establishments.

Third Meeting: Receive recommendations from members of the task force. Allow time for group feedback. Discuss how these recommended changes can be implemented. Remember to reward your task force for their work!

Follow-Up: Implement the solutions; evaluate, monitor, and modify where necessary.

2. Food-Safety Training Materials

Objective: *To assess current food-safety training practices for front-of-the-house personnel in your establishment.*

Directions: Most front-of-the-house personnel have received some orientation and training materials, but these materials may lack up-to-date information about adequate food safety and standards. Consider taking the following steps:

○ Assess all current training materials (videos, manuals, checklists, etc.) for the various front-of-the-house personnel. Are the appropriate food-safety issues addressed?

○ What additional safety issues and standards need to be addressed? How can these be integrated into existing materials?

○ How can you effectively communicate to training personnel the need for additional training in food-safety issues and standards? How can you be certain this information is reinforced during training?

○ Are employees properly tested about food-safety issues? Is there follow-up after the evaluation?

If you insist that employees follow safe foodhandling standards from the very beginning of their employment, and if you expect these standards to be met, and consistently monitor employee performance, employees will form safe foodhandling habits.

MULTIPLE-CHOICE STUDY QUESTIONS ???

1. Which serving method is most likely to protect the safety of the food being served?

   A. Using stainless-steel tongs to serve hot rolls directly to customers at the table

   B. Using a clean coffee cup to ladle soup into individual serving bowls

   C. Using a drinking glass to scoop ice cubes out of the icemaker storage compartment

   D. Using a teaspoon to scoop a serving of ice cream out of a five-gallon tub

2. Which procedure is the safest for dispensing ice cubes for beverage service?

   A. From an ice bucket with clean hands

   B. From the ice maker storage bin with a clean, sanitary metal or plastic scoop

   C. From the ice maker storage bin with a clean drinking glass

   D. From an ice bucket with a handled cup

3. Which bread product may be re-served to customers?

   A. Hard-crust rolls that were not eaten and are less than one day old

   B. Bread slices that were not eaten by the previous customer

   C. Bread sticks that are wrapped in cellophane

   D. Oyster crackers that were served in a bowl to the previous customer

4. Which of the following does not represent a potential food-safety problem for a self-service lunch buffet?

   A. A bowl of macaroni salad that has been sitting on a table at room temperature from 11:00 a.m. to 3:30 p.m.

   B. A chafing dish of chicken à la king served at a temperature of 120°F (49°C)

   C. Vegetable soup served from a large bowl

   D. A platter of fresh vegetables presented directly on ice

5. You are in charge of maintaining your restaurant's salad bar. Which of the following procedures is not necessary to prevent the customers from accidentally contaminating the food?

A. Provide a supply of clean drinking glasses every two hours at a minimum.

B. Give each customer a clean plate when he or she comes back for a second helping of food.

C. Locate the salad greens at the opposite end of the table from the roast ham.

D. Provide a proper serving utensil for every dish.

6. Which buffet setup practice demonstrates the FIFO principle?

A. Bowls of washed salad greens are put out for consumption one at a time.

B. Dated single-serve containers of milk are offered for consumption in order by date.

C. A bowl of carrot and raisin salad is refilled as needed throughout the day.

D. A basket of sliced bread is replaced every hour throughout the day.

7. You supply home-delivery boxed meals to elderly people in your community. The meals, one hot and one cold meal daily, are prepared in a hospital kitchen. The van driver takes you to the clients' homes in a specially equipped van. Which of the following conditions might give you reason to question the safety of the meals?

A. The temperature of the first cold meal delivered registered at 35°F (2°C).

B. The total time for the delivery route was more than two hours.

C. The driver didn't wash her hands before starting on the route.

D. The temperature of the last hot meal delivered registered at 135°F (57°C).

8. Hash brown potatoes are a popular item on your hotel's breakfast buffet. They are served in a chafing dish which has its own heat source and an external dial thermometer. How can you ensure that the potatoes are safe to eat?

A. Every two hours, measure the internal temperature of the potatoes with a thermometer.

B. Every two hours, record the temperature reading on the chafing dish's external thermometer.

C. Every hour, add fresh hot potatoes to the potatoes already in the chafing dish.

D. Every hour, turn up the heat on the chafing dish to 165°F (74°C).

9. Which of the following is an appropriate service of an uncooked food?

A. Raw fish in sushi, held for several hours and served at room temperature

B. Unshucked oysters on ice, partitioned off from other ready-to-eat foods

C. Raw chicken with fresh bread, served on a salad bar next to cut raw vegetables

D. Raw lamb kibbe, served on an appetizer tray with cold stuffed grape leaves, relishes, salads, and other ready-to-eat foods

Section 9
Principles of a HACCP System

Knowledge

TEST YOUR FOOD-SAFETY KNOWLEDGE

1. **True or False:** A control point (CP) is the last point where you can intervene to prevent, control, or eliminate the growth of microorganisms in food. *(See Determine Critical Control Points, page 9-6.)*

2. **True or False:** Cooking chicken to a minimum internal temperature of 165°F (74°C) for fifteen seconds would be an appropriate critical limit if the cooking step is a CCP for chicken. *(See Establish Critical Limits, page 9-7.)*

3. **True or False:** The first step in developing a HACCP system is to identify all Critical Control Points in the flow of food in your establishment. *(See Hazard Analysis, page 9-4.)*

4. **True or False:** An establishment's HACCP plan must be reevaluated if there is a menu change. *(See Verify that the System Works, page 9-10.)*

5. **True or False:** The receiving step would be a Critical Control Point when receiving fresh clams for clam chowder in an establishment. *(See Determine Critical Control Points, page 9-6.)*

Learning Essentials

After completing this section, you should be able to:

○ Identify the flow of a food through an establishment.

○ Discuss the importance of prerequisite programs for a Hazard Analysis Critical Control Point (HACCP) system.

○ Identify the HACCP principles for food safety.

○ Identify Critical Control Points (CCPs) for various foods and processes.

CONCEPTS

○ **Hazards:** Biological, chemical, or physical agents that may cause illness or injury if not controlled throughout the flow of food.

○ **Hazard Analysis Critical Control Point (HACCP):** A food-safety system designed to keep food safe throughout its flow in an establishment. HACCP is based on the idea that if hazards are identified at specific points in a food's flow, the hazards can be prevented, eliminated, or reduced to safe levels.

○ **Prerequisite programs:** The programs used in an establishment that protect your food from contamination, minimize microbial growth, and ensure the proper functioning of equipment.

○ **Hazard analysis:** The process of identifying and evaluating potential hazards associated with foods in order to decide which foods must be addressed in a HACCP plan.

○ **Control point (CP):** Any step in the flow of food where a physical, chemical, or biological hazard can be controlled.

○ **Critical Control Point (CCP):** The last step where you can intervene to prevent, control, or eliminate the growth of microorganisms before the food is served to customers.

○ **Critical limit:** Minimum and maximum limits that the CCP must meet in order to prevent, eliminate, or reduce a hazard to an acceptable limit.

○ **Monitoring:** The process of analyzing whether your critical limits are being met and you are doing things right.

○ **Corrective action:** A predetermined step taken when food doesn't meet a critical limit.

○ **Verification:** The step where you verify that the CCPs and critical limits you selected are appropriate, that monitoring alerts you to hazards, that corrective actions are adequate to prevent foodborne illness from occurring, and that employees are following established procedures.

INTRODUCTION

The last several sections talked about the flow of food through a typical establishment. In these chapters, you learned how to receive, store, prepare, cook, cool, reheat, and serve food.

In this section, we'll discuss a system that will enable you to consistently serve safe food by identifying and controlling possible hazards (biological, chemical, or physical agents that may cause illness or injury if not controlled) throughout the flow of food. This system is called Hazard Analysis Critical

Control Point (HACCP). A HACCP system is a dynamic process that uses a combination of proper foodhandling procedures, monitoring techniques, and record keeping to help ensure that the food you serve is safe. Because HACCP is dynamic, it allows you to continuously improve your food-safety system.

This section will provide you with a basic understanding of HACCP and its application to the industry; however, if you want to develop and implement a HACCP plan you will require additional information. A HACCP plan is a written document based on HACCP principles, which describes the procedures a particular establishment will follow.

WHAT IS HACCP?

The HAACP system is based on the idea that if biological, chemical, or physical hazards are identified at specific points within the flow of food, they can be prevented, eliminated, or reduced to safe levels. A HACCP system helps you to do the following.

○ Identify the foods and procedures that are most likely to cause foodborne illness.

○ Develop procedures that will reduce the risk of a foodborne-illness outbreak.

○ Monitor procedures to keep food safe.

○ Verify that the food you serve is consistently safe.

Prerequisite Programs

Prerequisite programs support your HACCP plan and are the basic operating conditions for producing safe food.

Prerequisite programs, or standard operating procedures (SOPs), protect your food from contamination, minimize microbial growth, and ensure the proper functioning of equipment. They may include the following.

○ Proper personal hygiene   ○ Creating supplier specifications

○ Proper facility design   ○ Proper cleaning and sanitation

○ Choosing good suppliers   ○ Appropriate equipment maintenance

DEVELOPING A HACCP PLAN

A HACCP plan is a written document that describes the procedures a particular establishment will follow. A HACCP plan is developed using the HACCP principles and is specific to the facility, its menu, its equipment, its processes, and its operations.

The HACCP plan for a product prepared in one facility will be different from the HACCP plan for the same product prepared in another facility.

While generic HACCP plans can serve as useful guides, each facility must develop a HACCP plan that addresses its own unique conditions.

HACCP Principles

The plan you develop will be based on the seven basic HACCP principles. Each principle builds upon the information gained from the previous principle.

○ Principles one, two, and three help you design your system.

○ Principles four and five help you implement it.

○ Principles six and seven help you maintain the system and verify its effectiveness.

Hazard Analysis (Principle 1)

To perform a hazard analysis in your establishment, you need to consider the following.

○ All the ingredients used in your menu items

○ Your equipment and processes

○ Your employees

○ The customers you serve

Here are some key steps to follow to identify all potential hazards in your establishment.

○ **Identify potential food hazards.** Identify any foods that may become contaminated if handled incorrectly at any stage in the flow of food, or that may allow the growth of harmful microorganisms. Make a list of these foods.

For example, your list may include menu items that contain chicken, seafood, pork, and beef, all of which are potentially hazardous as indicated in the illustration on this page.

○ **Determine where hazards can occur in the flow of food.** To do this, use these guidelines.

● Using your list, write down each step in the flow of food for each particular item.

Enrico's
Italian Dining

Appetizers:

Nachos Supreme

Crisp corn tortilla chips with our not-too-spicy ranchero sauce, topped with jack cheese, garden vegetables, and sour cream.

Grilled Vegetable Quesadilla

Our quesadilla filled with green peppers, red onion, zucchini, yellow squash, jack and cheddar cheese. Served with guacamole, sour cream, salsa, and garnished with honey mustard and BBQ sauce.

Specialties of the house:

CHICKEN BREAST alla PARMAGIANA

Boneless, skinless breast of chicken seasoned with herbs, lightly breaded & pan-fried. Baked with three cheeses & herb marinara sauce.

ENRICO S PEPPER STEAK

Beef tenderloin medallions sauteed in butter with fresh mushrooms, sweet bell peppers, and onions.

BAY SCALLOPS

Sauteed with fresh mushrooms, served over fettucine alla Enrico.

PORK TENDERLOIN alla MARSALA

Medallions of pork sauteed in butter with fresh mushrooms and sweet Marsala wine.

Potentially Hazardous Foods List

○ **Cheese**
○ **Sour Cream**
○ **Guacamole**
○ **Chicken**
○ **Beef**
○ **Bay Scallops**
○ **Pork**

Identifying Potentially Hazardous Foods on a Menu

Examine and write down all the potentially adverse conditions the food may be exposed to during its flow through the establishment. Consider how your operation's equipment may affect these foods. Determine the potential hazards that could occur, such as contamination and time-temperature abuse, and write them down.

In our example, chicken was identified as a potentially hazardous food. We determined and recorded its flow, which includes receiving, storage, preparation, cooking, and service. There are potentially adverse conditions at each step. In the illustration below, potentially adverse conditions have been identified for the preparation step, but these conditions must be identified for all steps in your establishment.

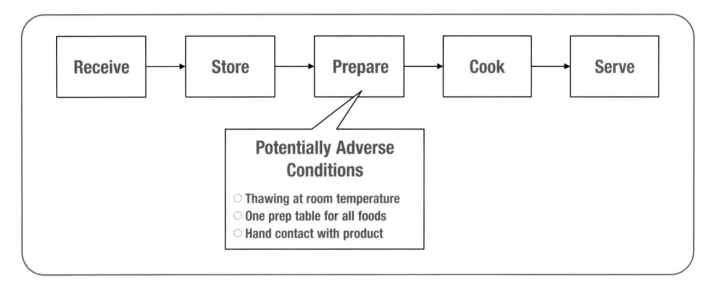

Determining the Steps in the Flow of Chicken Through an Establishment

○ **Group foods by how they are processed.** The most common groups include:

● Foods that are prepared and served without cooking (salads, raw oysters, cheeses, and sandwich meats). Hazards involve contamination of the food by employees or equipment.

● Foods that are prepared and cooked for immediate service (hamburgers, scrambled eggs, and hot sandwiches). Hazards involve improper cooking, which may not eliminate biological hazards.

● Foods that will be prepared, cooked, held, cooled, reheated, and served (chili, soups, and sauces). Hazards may occur at many points.

The chicken in our example is prepared and cooked for immediate service. Pork, which is also found on the list of potentially hazardous menu items, is processed in this establishment in the same manner, and therefore can be grouped with chicken. Since beef is used only for minestrone soup in this establishment, it falls under the third grouping of processes.

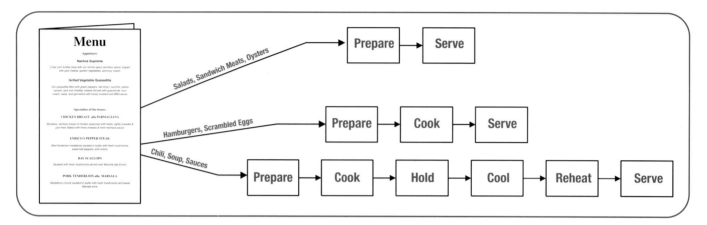

Grouping Foods by Processes

When developing your HACCP plan, you may choose to group foods by how they are processed in your establishment.

○ **Identify your customers.** This is particularly important if the customers you serve are very young or elderly, or people who are ill or immunocompromised.

Determine Critical Control Points (Principle 2)

Once you've identified all potential food hazards and the step or steps at which they occur in your establishment, the next task is determining at which steps we can intervene to control these hazards. To make this determination, consider the following guidelines.

○ Find any step in the flow of food where a physical, chemical, or biological hazard can be controlled; this is a control point.

○ To assess whether a control point is critical, you need to determine if it is the last step where you can intervene to prevent, control, or eliminate the growth of microorganisms before the food is served to customers. If it is, the step is called a Critical Control Point (CCP).

○ Cooking, chilling, or holding are typically CCPs. However, these may not be the CCPs for all foods or all processes in your establishment.

We know that the raw chicken in our example may be contaminated with the biological hazards *Salmonella* and *Campylobacter*. While care is needed during preparation to prevent cross-contamination, proper cooking is essential to prevent your customers from becoming ill. Therefore, preparation is a control point, while cooking is a Critical Control Point for this menu item prepared with raw chicken, as seen on the next page.

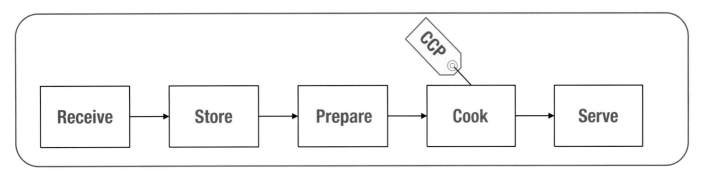

Identifying Critical Control Points in the Flow of Chicken
Since cooking is the last step where biological hazards can be prevented, controlled, or eliminated for the chicken in this establishment, this step is a Critical Control Point (CCP).

Establish Critical Limits (Principle 3)

Once you've determined the CCPs for the potential food hazards in the flow of food, you need to establish critical limits, which are minimum and maximum limits that the CCP must meet in order to prevent, eliminate, or reduce a hazard to an acceptable limit. When establishing critical limits, keep the following points in mind. The limit must be:

○ Measurable (such as a time or a temperature).

○ Based on scientific data, food regulations (such as the FDA Model Food Code), and expert advice.

○ Appropriate for the food and equipment when prepared under normal conditions and specific to your establishment.

○ Clear and easy to follow.

For example, in our establishment, we have determined that cooking is a CCP for chicken, and we have set our critical limit for cooking to a minimum of 165°F (74°C) for fifteen seconds.

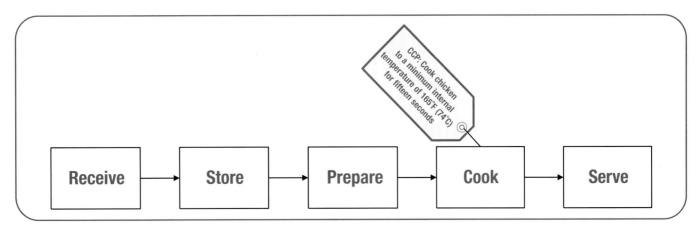

Establishing Critical Limits for Cooking Chicken

Monitoring Critical Control Points (Principle 4)

Monitoring lets you know that critical limits are being met, and that you are doing things right.

To develop a successful monitoring program, you need to focus on each CCP, and establish clear directions that will determine the following.

○ **How to monitor the CCP.** This depends on the critical limits you have established and may include measuring time, temperature, pH, oxygen, water activity, etc.

○ **When and how often to monitor the CCP.** Continuous monitoring is preferable but not always possible. Regular monitoring intervals should be determined based on the normal working conditions in your establishment, depending on volume, equipment, and procedures.

○ **Who will monitor the CCP.** Assign responsibility to a specific employee or position and make sure that person is trained to perform the monitoring.

○ **Equipment, materials, or tools needed to monitor the CCP.** Provide employees with the proper items to make the task as easy as possible.

Going back to the example of the chicken, we have established (1) that cooking is a CCP; and (2) the critical limits are 165°F (74°C) for fifteen seconds. Because our facility cooks chicken to order, we can take the temperature of each chicken at this point. We determined that our cook needs to monitor the chicken to see that it has reached the proper temperature for the required amount of time (reached the critical limit) by inserting a thermometer into several places in the thickest part of each chicken.

Monitoring the CCP

Cooks must use a thermometer to verify that each chicken portion has reached a minimum internal temperature of 165°F (74°C) for fifteen seconds. Take at least two readings in different locations. Insert the thermometer in the thickest part of each chicken portion.

CCP: Cook chicken to a minimum internal temperature of 165°F (74°C) for fifteen seconds

Cook

Monitoring the CCP for Chicken

Taking Corrective Action (Principle 5)

Corrective actions are predetermined steps taken when food doesn't meet a critical limit. Remember, this will be the last opportunity you have to ensure the safety of the food served. Corrective actions may include the following.

○ Continuing to cook the food to the correct temperature

○ Throwing food away after a specified amount of time

○ Rejecting a shipment that is not received at the temperature you specified

When developing corrective actions:

○ Be sure that the action you name is specific. It should mention exactly what should be done to correct the situation.

○ Employees must know who is responsible for taking a corrective action, how to perform the action, and where and how to record the action that was taken.

In our previous example, the manager determined that if cooked chicken doesn't reach its critical limit, the corrective action is that the cook should continue cooking the chicken until it reaches that limit. The cook is also instructed to record this corrective action in the log.

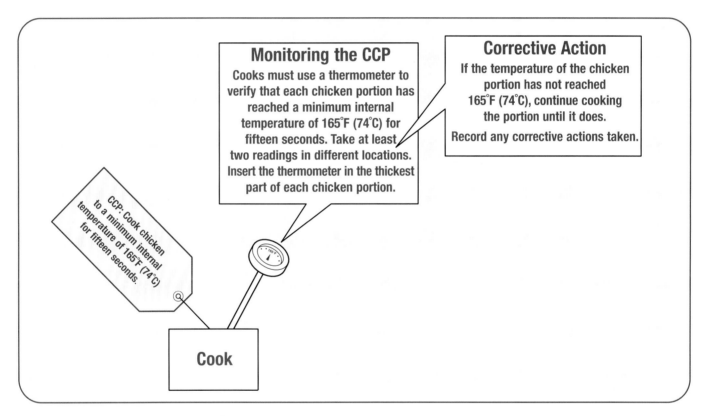

Monitoring the CCP
Cooks must use a thermometer to verify that each chicken portion has reached a minimum internal temperature of 165°F (74°C) for fifteen seconds. Take at least two readings in different locations. Insert the thermometer in the thickest part of each chicken portion.

Corrective Action
If the temperature of the chicken portion has not reached 165°F (74°C), continue cooking the portion until it does.
Record any corrective actions taken.

CCP: Cook chicken to a minimum internal temperature of 165°F (74°C) for fifteen seconds.

Cook

Corrective Action for Cooking Chicken

Verify that the System Works (Principle 6)

After you have developed your system, you need to confirm that it works according to the plan. This is called verification. During this step you should verify the following.

○ CCPs and critical limits you selected are appropriate.

○ Monitoring alerts you to hazards.

○ Corrective actions are adequate to prevent foodborne illness from occurring.

○ Employees are following established procedures.

Verification of your plan should be performed on a regular basis. Reevaluate your HACCP plan if:

○ You notice critical limits are frequently not being met.

○ You receive a foodborne-illness complaint.

○ There are any changes in your menu, equipment, processes, suppliers, or product.

In our example, the six-month plan evaluation revealed that we frequently failed to meet the critical limits set for our chicken. Upon reevaluating our cooking process, we determined that our supplier had started providing a larger product. This caused the chicken to be undercooked, given the equipment and cooking parameters we had established. We revised our plan by adjusting our cooking process to account for larger weights.

Record Keeping and Documentation (Principle 7)

Recording how food is produced and kept safe is important to the success of a HACCP system. Proper records allow you to:

○ Document that you are continuously preparing and serving safe food.

○ Identify when your process should be modified due to food-safety problems that have been noted.

Examples of records include time-temperature logs, procedures for taking temperatures, SOPs, calibration records, corrective actions, monitoring schedules, product specs, etc.

SUMMARY

Hazard Analysis Critical Control Point (HACCP) is a food-safety system designed to keep food safe throughout its flow in an establishment. HAACP is based on the idea that if biological, chemical, or physical hazards are identified at specific points within a food's flow through the operation, the hazards can be prevented, eliminated, or reduced to safe levels. A successful HACCP system uses a combination of proper foodhandling procedures, monitoring techniques, and record keeping to keep food safe.

HACCP must be built on a solid foundation of prerequisite programs. These programs protect your food from contamination, minimize microbial growth, and ensure the proper functioning of equipment. They include programs for proper personal hygiene, proper cleaning and sanitation, proper facility design, choosing good suppliers, and equipment maintenance.

While generic HACCP plans can serve as useful guides, each facility must develop a plan that addresses its own unique conditions. The plan should be specific to the facility, its menu, its equipment, its processes, and its operations. An effective HACCP plan will be based on the following seven basic HACCP principles.

- Conduct a hazard analysis.
- Determine the Critical Control Points (CCPs).
- Establish critical limits.
- Establish monitoring procedures.
- Establish corrective actions.
- Establish verification procedures.
- Establish record-keeping and documentation procedures.

NOTES

ACTIVITY

Crossword Puzzle

Across:

3. Predetermined steps taken when food doesn't meet a critical limit.

4. Minimum and maximum limit that the CCP must meet.

7. The last step in a food's flow where you can intervene to keep the food safe.

8. Biological, chemical, or physical agents that may cause illness.

9. The step where you confirm that the CCPs and critical limits you selected are appropriate.

10. Basic operating practices for producing safe food. (Abbreviation)

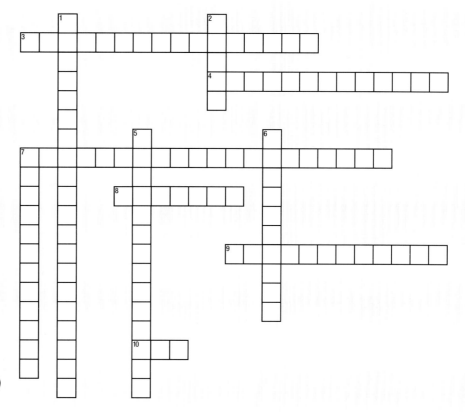

Down:

1. Proper personal hygiene, proper facility design, and choosing good suppliers are examples of these types of programs.

2. This system is based on the idea of preventing, eliminating, or reducing hazards to safe levels. (Abbreviation)

5. The process of identifying and evaluating potential hazards.

6. This process lets you know if you are doing things right.

7. Any step in the flow of food where a hazard can be controlled.

ACTIVITY

Word Find

Find the terms that go with the clues below.

Clues:

1. Proper personal hygiene, proper facility design, and choosing good suppliers are examples of these types of programs.

2. This system is based on the idea of preventing, eliminating, or reducing hazards to safe levels. (Abbreviation)

3. The process of identifying and evaluating potential hazards.

4. Basic operating practices for producing safe food. (Abbreviation)

5. Any step in the flow of food where a hazard can be controlled.

6. Minimum and maximum limit that the CCP must meet.

```
C O R R E C T I V E A C T I O N M C W E F M A O S E
Z V J C Z T O P P O E T N I O P L O R T N O C I R P
E E Z D C J K O N W X O G S V B Z A F G V P S U C R
B K C H E V C S O V I W D S P T W J M Q W Y D Q R E
X Y V X O V S A Z D W P O K X Y M Z C Z L E R E I R
V I L V P P C I L D L N R G T V I K S A C J H P T E
K S C J R M Y N B I H X X F Q E Q E N O I E S W I Q
B C R R T A B U D U E P Z W V F C A R X C Z O P C U
M I J P I T A D G I C P O N N B D P X M M I R V A I
B N A X O T P Z G L G W Q R O R G X O O W D H W L S
K M V Z B M I C Q C B X Q A A N C O I G L Q V V L I
S J Q W X Z D C C V G U W Z I M Z C K Q Z N D D I T
N Q G C U S N R A A M B A T J D V R T A K F Q R M E
O S L K E U H W H L H H A L C Q V Z C V V J M G I P
I U J C R E U L R S C R W Y L A S X F D X A N R T R
T Q Q R L T L R I E E O L T P W P E D N Q W S C G O
A K B P E L K P H P Q Q N G Q V F J U O K Z Y T X G
C P F O J X H U O Q F D J T N H A A P K S S N K B R
I J I K C H S D V Y F R K N R I S W M F C K T C L A
F O O X A D R Y Q F O N Z C Y O R L X N L K Q K D M
I T I Z R A N Y I W R P K O E L L O K G K O H C K S
R G Q A D V M S O A H X D W T Q U P T X L L K F P V
E F Z N R I C O F U B Y W F A W A L O I Z L O O V X
V A A I F Q J F P T U M H D B M G I L I N X M Y X H
H T S O M J T L Q N T I Z N C W F U L M N O A K R H
S Z E L W Q D Y Z Y M C G N A P T T S H Y T M G C A
```

7. This process lets you know if you are doing things right.

8. Biological, chemical, or physical agents that may cause illness.

9. The last step in a food's flow where you can intervene to keep the food safe.

10. The step where you confirm that the CCPs and critical limits you selected are appropriate.

11. Predetermined steps taken when food doesn't meet a critical limit.

ACTIVITY

Let's Take a Look at the Menu!

From this section, you have learned that grouping foods by process may help you identify the potential hazards in your operation. Based on your own menu, group foods using the processes below as a guide. Be able to explain the rationale behind your groupings.

- **Process 1:** Foods that are prepared and served without cooking. Hazards involve contamination of food by employees or equipment.

- **Process 2:** Foods that are prepared, cooked, and served. Hazards involve improper cooking, which may not eliminate biological hazards.

- **Process 3:** Foods that will be prepared, cooked, held, cooled, reheated, and served. Hazards may occur at many points.

ACTIVITY

Meet the Critical Control Cop

Below you will find an illustration of each of the seven HACCP principles. Identify the principle and give a short explanation of each.

1.

2.

3.

4.
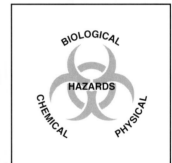

5.

6.

7.

A CASE IN POINT

Case Study

Jason, the manager at Cal's Catering, received several calls from customers complaining that they had contracted a foodborne illness after attending a picnic catered by Cal's the previous weekend. Concerned, Jason reviewed his menu for the event and noticed that the only potentially hazardous food served was barbecued rib sandwiches. He looked at his HACCP plan and noticed that the barbecued rib sandwiches were received, stored, prepared, and cooked in the establishment. They were put in large, insulated containers and reheated and held for self-service during the picnic.

At what steps in the flow of the barbecued rib sandwiches should Jason have identified Critical Control Points? What critical limits should have been established for each CCP? What records should be available for this plan?

NOTES

TRAINING TIPS

Training Tips for the Classroom

1. Developing HACCP Systems—A Breakout Exercise

Objective: *After completing this activity, class participants should be able to assess hazards and identify Critical Control Points (CCPs) for various menu items.*

Directions: Make a list of four to eight different menu items. Identify the method of preparation for each item and the type of establishment that might serve it. Print up handouts with one menu item per page. Examples may include:

Menu Item	Method of Preparation	Establishment
Chili	Prepare, cook, hold, cool, reheat	Full service
Potato Salad	Prepare, cook, cool, hold	Vending machine
Baked Chicken	Prepare, cook, hold, serve	Catered event
Cream Pie	Prepare, serve	Retail

When designing this list, include different types of foods (meats, fish, dairy, vegetables, etc.). Include potentially hazardous foods (PHFs) or items that contain PHFs. Also, vary the methods of preparation (cook, cool and reheat; prepare and serve; hot hold), and assign these menu items to different types of establishments (quick service, full service, catering, vending, retail, and so on). When developing this list, keep your audience in mind. If possible, the list should reflect the types of food and the types of establishments that your students represent.

Break your class into four to eight teams, with two to four members each. Assign a captain for each team. (Choose team captains who have some foodservice experience.) Assign each team a menu item. If the captain comes from a specific type of establishment, assign that team a menu item produced in a similar establishment.

Give each team a handout with the seven HACCP principles listed on it. Ask each team to develop a HACCP plan for their assigned menu item within the specified type of foodservice operation. It is not as important that each group develop a complete plan as it is that they properly assess the hazards associated with their menu item and identify the Critical Control Points (CCPs).

This exercise will last about an hour. Allow thirty minutes for the teams to develop the outline for the HACCP plan, and thirty minutes for the teams to present their plans to the class. You should also allow adequate time for questions and comments from the class.

2. Control Point (CP) or Critical Control Point (CCP)?

Objective: *After completing this activity, class participants should be able to differentiate between control points (CPs) and Critical Control Points (CCPs).*

Directions: Develop a list of ten food items along with a specific step or process related to the preparation and service of that particular item. Here are some examples (with answers in bold):

Item	Step or Process
Fresh chicken	Receive chicken at 41°F (5°C) or below. **(CP)**
Fresh ground beef	Discard ground beef that has been in the temperature danger zone for more than four hours. **(CP)**
Fresh pork	Cook pork to a minimum internal temperature of 155°F (68°C). **(CCP)**
Iceberg lettuce	Wash lettuce prior to making salads. **(CP)**
Chili	Hold cooked chili for service at 140°F (60°C). **(CCP)**
Clam chowder	Cool cooked clam chowder to 70°F (21°C) within two hours and to 41°F (5°C) within an additional four hours. **(CCP)**

Give a copy of the list to each student. Ask them to determine whether the step or process for each food in the list represents a Control Point or Critical Control Point.

As a group, discuss each menu item. Discuss rationales for each student's position on CPs and CCPs. If possible, reach an agreement about each item.

3. The Seven Principles Pop Quiz

Objective: *After completing this activity, class participants should be able to explain each of the seven HACCP principles and give an example of each.*

Directions: After discussing Section 9, have your students close their books and put away their notes. Give the students a blank sheet of paper and ask them to write down, in order, the seven principles for setting up a HACCP system. They should explain what each principle means and give a specific example.

Allow fifteen minutes for this quiz. Then review the seven principles, and provide students with a completed, one-page synopsis of the seven principles to use as a study aid.

Training Tips on the Job

1. *Developing and Implementing a HACCP Plan*

Purpose: *To provide some practical tips to help organize the development and implementation of a HACCP plan.*

Directions: Developing a HACCP plan can take a fair amount of time and effort, so it may be more effective if developed and implemented in several stages. Here are a few suggestions.

Stage 1. First, make sure that your establishment has strong prerequisite programs in place. These include programs for personal hygiene, cleaning and sanitation, equipment maintenance, supplier selection, and so on. Strong prerequisite programs are the foundation of a strong food-safety system, and are essential to the success of any HACCP plan.

Stage 2. Build a HACCP team. Ideally, the team should represent people from different job positions who handle some aspect of food production and service. For example, the team might include an employee responsible for ordering, receiving, and storing foods; a chef and a prep cook; a server; and a manager, among others. Team members should have food-safety knowledge and strong foodservice experience. They should also be respected by their co-workers, and be committed to the program.

Stage 3. The members of the team should meet to discuss the food served in your establishment, how it is handled, the nature of your customers, and the prerequisite programs currently in place.

Stage 4. After the initial discussion, the team will address the seven HACCP principles. This stage should probably be broken down into several meetings. Keep in mind that all seven principles must be performed in order, because each principle builds on the previous one. The seven principles can be combined into three groups to aid in development: Principles one, two, and three (Stage 4); principles four and five (Stage 5); and principles six and seven. The team must begin by conducting a hazard analysis. A brainstorming session may help the team identify a list of potential hazards likely to occur in the flow of food in the establishment (hazard analysis, Principle 1). Each team member brings to the table potential hazards specific to their area in the establishment. The team should review ingredients used in menu items, processes for preparing the items, equipment used to process it, and how the product will be stored and distributed.

After the list has been completed, the team must then determine which potential hazards will be addressed in the HACCP plan (determining Critical Control Points, Principle 2). The choice should be based on the severity of the hazard and the likelihood that it will occur in the establishment. These are the Critical Control Points (CCPs) that must be identified by the team. Once the CCPs are identified, the team should brainstorm methods for controlling them (setting critical limits, Principle 3). The controls will be critical to preventing, eliminating, or reducing the hazard to an acceptable level. If necessary, the team should look to regulatory standards and guidelines or experts to help set the critical limits.

Stage 5. Next, the team must establish procedures for monitoring the CCPs (Principle 4), and corrective actions to take to keep the CCPs in control (Principle 5). Make sure these procedures are specific so that the employee responsible for performing the monitoring or corrective action knows exactly what is expected. Training is the key to a successful implementation.

Stage 6. A HACCP plan should be implemented gradually over a period of time in order to be successful. Furthermore, you must frequently revisit and reassess your HACCP plan, particularly when changes in your establishment take place (this is Principle 6, verification). Changes that might affect your plan could include employee turnover, a change in menu items and products, changes in equipment, changes in company standards, or new local or state laws.

Be sure your team establishes sound record-keeping procedures. Records will keep an accurate account of the who, what, when, where, and why for monitoring and corrective actions, and will also aid in verifying your plan.

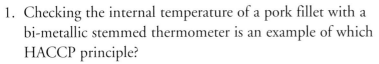

MULTIPLE-CHOICE STUDY QUESTIONS

1. Checking the internal temperature of a pork fillet with a bi-metallic stemmed thermometer is an example of which HACCP principle?

   A. Verification   C. Record keeping

   B. Monitoring   D. Hazard analysis

2. You place chicken salad from a buffet bar in an ice-water bath to cool after you find that its temperature is 55°F (13°C) instead of 41°F (5°C) or below. This is an example of

   A. monitoring.   C. a hazard analysis.

   B. corrective action.   D. verification.

3. Your restaurant plans to add linguini in a red clam sauce to the menu. The fresh clams will be cooked before being mixed with the pasta. Should receiving be a CCP for the clams?

   A. Yes, receiving is always a CCP for clams.

   B. No, clams do not pose a health hazard.

   C. No, because the clams will be cooked thoroughly.

   D. Yes, because the clams will be cooked thoroughly.

4. All of the following are methods of monitoring a CCP except

   A. taking the temperature of food during the cooking process.

   B. observing the amount of time food remains in the temperature danger zone.

   C. checking the pH of a food.

   D. analyzing a food's flow through the establishment.

5. Which of the following would not be a corrective action?

   A. Continuing to cook a hamburger that has not reached an internal temperature of 155°F (68°C)

   B. Throwing out potato salad that has remained at room temperature for longer than four hours

   C. Covering a cut with a bandage and finger cot

   D. Rejecting a delivery of fish received at an internal temperature of 60°F (16°C)

6. Which statement best describes the purpose of verification in a HACCP plan?

   A. To determine if the CCPs and critical limits that have been chosen are appropriate

   B. To determine whether monitoring is alerting you to hazards

   C. To determine if corrective actions are adequate to prevent foodborne illness

   D. All of the above

7. Which of the following records would not be useful to your HACCP plan?

   A. Time and temperature logs

   B. Thermometer calibration records

   C. Corrective action logs

   D. Workplace accident records

8. Which of the following prerequisite programs should be included in your HACCP Plan?

   A. A personal hygiene program

   B. An incentive program

   C. Workplace accident prevention program

   D. None of the above

9. Which of the following situations could sabotage even the best HACCP plan at an establishment?

   A. Improper procedures for cooling food

   B. Faulty equipment that won't maintain temperatures

   C. Ineffective handwashing practices

   D. All of the above

10. The purpose of a HACCP system is to

    A. identify and control possible hazards throughout the flow of food.

    B. identify the proper methods for receiving foods.

    C. keep the establishment pest free.

    D. identify faulty equipment within the establishment.

11. Which of the following steps is likely to be a Critical Control Point for oysters that will be eaten raw?

    A. Receiving
    B. Storage
    C. Preparation
    D. All of the above

12. A chef took the temperature of a container of minestrone soup which was being held in a hot-holding unit for service. The temperature of the soup was 120°F (49°C), which he recorded in a temperature log. The chef reheated the soup to 165°F (74°C) for fifteen seconds and placed it in a different holding unit. Which of the following actions was a corrective action?

    A. Taking the temperature of the soup
    B. Reheating the soup
    C. Recording the temperature of the soup in the temperature log
    D. Placing the soup in a different holding unit

13. Which of the following statements is an example of a critical limit?

    A. Cook ground beef to 155°F (68°C).
    B. Store ground beef at 41°F (5°C).
    C. Discard ground beef if it remains at temperatures between 41°F and 140°F (5°C and 60°C) for more than four hours.
    D. All of the above

14. Your deli serves cold sandwiches in a grab-and-go display. Which step in the preparation of these sandwiches may be a CCP?

    A. Storage   C. Reheating
    B. Cooking   D. Cooling

15. Noting food temperatures in a temperature log is an example of which HACCP principle?

    A. Verification   C. Record keeping
    B. Monitoring   D. Hazard analysis

UNIT 3

MANAGING YOUR OPERATION

At Taco Bell, we are extremely dedicated to food safety. Serving 50 million customers weekly imposes an enormous responsibility to keep them safe. Ninety-percent of our managers are ServSafe certified. Food safety is one of those things you've got to get right the first time. You really only have one shot.

Gary DuBois
Director of Quality Assurance
And Food Safety
Taco Bell

Section 10
Sanitary Facilities and Pest Management

Knowledge

TEST YOUR FOOD-SAFETY KNOWLEDGE

1. **True or False:** A hose attached to a sink faucet and left sitting in a bucket of dirty water could contaminate the water supply. *(See Cross-Connections, page 10-10.)*

2. **True or False:** It is all right to use nonpotable water for scrubbing pots and pans. *(See Water Supply, page 10-9.)*

3. **True or False:** Chemicals may be stored in food-prep areas if the containers are properly labeled. *(See Cleaning Tools and Supplies, page 10-20.)*

4. **True or False:** A surface that has been cleaned is safe to use for food preparation. *(See Cleaning and Sanitizing, page 10-12.)*

5. **True or False:** Pesticides can be stored in food storage areas if they are properly labeled and closed tightly. *(See Using and Storing Pesticides, page 10-27.)*

Learning Essentials

After completing this section, you should be able to:

○ Respond to an interruption in the internal water supply.

○ Respond to wastewater overflows.

○ Identify methods to prevent backflow problems.

○ Compare and contrast the uses of potable and nonpotable water.

○ Identify the procedures for proper internal and external waste disposal.

○ Clean and maintain restrooms properly.

○ Identify the requirements of a handwashing station.

○ Identify the optimum placements for sanitation of equipment and facilities.

○ Explain storage requirements for pesticides, chemicals, and cleaning agents.

○ Identify cleaning and sanitizing procedures for equipment, utensils, and food-contact surfaces.

○ Identify factors that affect the efficiency of sanitizers.

○ List proper machine warewashing procedures.

○ Identify ways to prevent pests from entering the facility.

○ Detect signs of infestation and identify pests involved.

○ List possible control measures for pest infestation.

CONCEPTS

○ **Air gap:** An air space separating a water supply outlet from any potentially contaminated source. An example is the vertical distance between the rim of a sink and the faucet. An air gap is the only completely reliable method for preventing backflow.

○ **Backflow:** The unwanted reverse flow of contaminants through a cross-connection into a potable water system. It occurs when the pressure in the potable water supply drops below the pressure of the contaminated supply.

○ **Cross-connection:** A physical link through which contaminants from drains, sewers, or wastewater can enter a potable water supply. A hose connected to water supply and submerged in a mop bucket is an example.

○ **Blast chiller:** Equipment designed to move food through the temperature danger zone quickly. Many are able to cool food from 140°F to 37°F (60°C to 3°C) within ninety minutes.

○ **Garbage:** Wet waste material, usually containing food, that cannot be recycled. It attracts pests and has the potential to contaminate food items, equipment, and utensils.

○ **NSF International:** Formerly known as the National Sanitation Foundation, NSF International publishes standards for foodservice equipment design.

○ **Potable water:** Water that is safe to drink or use as an ingredient in food.

○ **Single-use item:** Any article designed to be used once and thrown away. It includes items such as paper towels, plastic gloves, paper plates, plastic eating utensils, and single-serving condiment packets.

○ **UL (Underwriters Laboratories):** Provides sanitation classification listings for equipment found in compliance with NSF International standards.

○ **Vacuum breaker:** A device for preventing backflow between two water systems.

○ **Cleaning:** The process of removing food and other types of soil from a surface such as a countertop or plate.

○ **Sanitizing:** The process of reducing the number of microorganisms on a surface to safe levels.

○ **Heat sanitizing:** The most common way to heat sanitize tableware, utensils, or equipment is to submerge or spray these items with hot water.

○ **Chemical sanitizing:** The process of subjecting a food-contact surface to a sanitizing solution for a specific period of time to kill or reduce the number of microorganisms on that surface.

○ **Sanitizer:** A chemical used to sanitize. Chlorine, iodine, and quats are types of chemical sanitizers.

○ **Chlorine:** The most commonly used sanitizer due to its low cost, low corrosiveness, and effectiveness at low temperatures. Very effective against many microorganisms.

○ **Iodine:** A chemical sanitizer considered to be one of the most effective. Less corrosive and irritating to skin than chlorine.

○ **Quaternary Ammonium Compounds (quats):** A group of sanitizers all having the same basic chemical structure. Quats are generally nontoxic, noncorrosive, and stable when exposed to heat but may not kill certain types of microorganisms.

○ **Material Safety Data Sheet (MSDS):** A sheet that lists the chemical and common name of the material, potential physical and health hazards, and directions for safe handling and use.

○ **Master cleaning schedule:** A detailed schedule that lists all cleaning tasks in an establishment, when and how they are to be performed, and who will do the cleaning.

○ **Integrated Pest Management:** A system of preventive measures and control measures to prevent pests from entering your facility, and to eliminate existing pest infestations.

○ **Pest Control Operator (PCO):** A licensed or certified technician who implements and monitors control programs for companies that contract for services.

○ **Infestation:** The situation when pests overrun or inhabit an establishment in large numbers.

INTRODUCTION

When designing a sanitary facility you must consider how every area of it will be kept clean. The building and all equipment must be easy for employees to clean. Areas and equipment that are not properly cleaned allow bacteria, viruses, and molds to remain. These can become serious problems when food comes into contact with them. Facilities should be arranged so contact with contaminated sources such as garbage or dirty tableware, utensils, and equipment is unlikely to occur. Consider the following when designing an establishment.

○ Select materials for walls, floors, and ceilings that will make the surfaces easier to clean.

○ Arrange equipment and fixtures to comply with local regulatory standards.

○ Design the layout of utilities to prevent contamination and make cleaning easier.

○ Manage solid waste properly to avoid contaminating food and attracting pests.

The Plan Review

A sanitary layout and design should begin in the planning stage of the facility. Use the following guidelines when planning a facility.

○ Consult local regulations. Many jurisdictions require plans for new construction or extensive remodeling to be approved by the local health department or regulatory agency. Even if local regulations do not require it, it's a good idea to have plans reviewed by the health department anyway.

○ Depending on local regulations, local building and zoning departments will probably need to review and approve the plans too.

○ Be sure handicapped patrons and employees have access to the building as required by the Americans with Disabilities Act (ADA).

○ When construction is complete, the facility must obtain an operating permit from the local health department. This usually requires an inspection by the department to make sure all plans and regulations have been followed during construction.

Materials for Interior Construction

The most important consideration when selecting construction materials is how easy the establishment will be to clean and maintain.

There are advantages and disadvantages of different construction materials for floors, walls, and ceilings when constructing a new facility or simply remodeling an existing one. In this stage of planning, consult with experts in these areas.

CONSIDERATIONS FOR SPECIFIC AREAS OF THE FACILITY

Dry Storage

A well-designed dry storage area should be easy to clean and promote good air circulation. Consider the following guidelines.

○ Shelves, table surfaces, and bins should be made of corrosion-resistant metal, such as stainless steel, or food-grade plastic.

○ The area should be free of exposed steam pipes or heating ducts. They can raise the temperature of the room enough to affect the food.

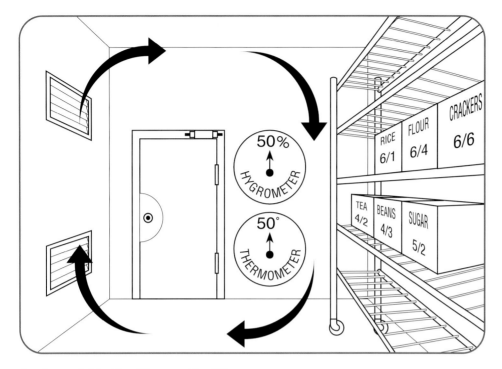

An Acceptable Dry Storage Facility
Dry storerooms should be constructed of easy-to-clean materials that allow good air circulation.

○ The area should be free of exposed water or sanitation pipes. Leaks or moisture from dripping condensation can promote microbial growth.

○ Outside windows and doors must have screens that are sixteen mesh to the inch, and all cracks in walls and floors must be filled to keep pests out.

Restrooms

Ideally, your establishment should have separate restroom facilities for customers and employees. If this is not practical, customers must not pass through food-preparation areas on their way to the restroom, because they could contaminate food or food-contact surfaces. Restrooms must have:

○ Fully equipped handwashing stations and self-closing doors.

○ Adequate stock. They should be stocked with toilet paper, and trash receptacles must be provided if disposable towels are used. Covered waste containers must be provided in women's restrooms for the disposal of sanitary supplies.

○ Regular cleaning. They should be cleaned from top to bottom at least once a day.

Handwashing Stations

Handwashing stations must be conveniently located so that employees will be encouraged to wash their hands often. These stations must be operable, stocked, and maintained. They are required in food-preparation areas, service areas, equipment-washing areas, and restrooms. A handwashing station must be equipped with the following items.

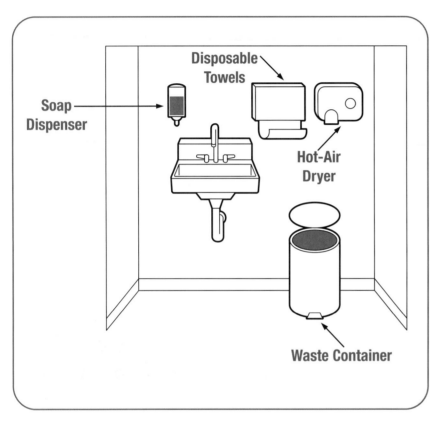

An Acceptable Handwashing Station
A handwashing station must be equipped with hot and cold running water, soap, a means to dry hands, and a waste container (if disposable towels are used).

○ **Hot and cold running water.** A water temperature of at least 110°F (43°C), supplied through a mixing valve or combination faucet, is required.

○ **Soap.** The soap may be liquid, bar, or powder. Liquid soap is generally preferred, and some local codes require liquid soap.

○ **A means to dry hands.** Most local codes require establishments to supply disposable paper towels in hand-washing stations. Continuous-cloth towel systems, if allowed, should be used only if the unit is working properly and the towel rolls are checked and changed regularly. Installing at least one hot-air dryer in a handwashing station may provide an alternate method for drying hands if paper or cloth towels run out. The use of common cloth towels is not permitted because they can transmit contamination from one person's hands to another.

○ **A waste container.** Waste containers are required if disposable paper towels are provided.

SANITATION STANDARDS FOR EQUIPMENT

The task of choosing equipment designed for sanitation has been simplified by organizations such as NSF International, formerly the National Sanitation Foundation. NSF International develops and publishes standards for sanitary equipment design. Underwriters Laboratories (UL) also provides sanitation classification listings for equipment found in compliance with NSF International standards. Foodservice managers should look for the NSF International mark or the UL Sanitation Classification mark on commercial foodservice equipment. To the right are examples of the NSF International and UL Sanitation Classification marks. The regular UL listing mark indicates compliance of equipment to UL safety standards.

Clean-in-Place Equipment

Some equipment, such as ice machines or soft-serve ice cream or frozen yogurt machines, are designed to be cleaned in place by flushing detergent, hot water, and sanitizing solution through it. This process should be done daily unless otherwise indicated by the manufacturer. Cleaning and sanitizing solutions must:

○ Remain within a fixed system of pipes for a predetermined amount of time.

○ Not leak into the rest of the machine.

○ Reach all food-contact surfaces.

Refrigerators and Freezers

When purchasing a refrigerator or freezer, make sure that it is NSF International certified or UL classified for sanitation (or the equivalent).

Blast Chillers and Tumble Chillers

Blast chillers are designed to move food through the temperature danger zone quickly. Many blast chillers are able to cool food from 140°F to 37°F (60°C to 3°C) within ninety minutes. Once chilled to safe temperatures, the food can then be stored in conventional refrigerators or freezers. Tumble chillers are also designed to cool food quickly.

Cutting Boards

Cutting boards are an important part of the establishment. Consider the following points when selecting a cutting board.

○ Synthetic cutting boards are often preferred over wooden boards because they can be cleaned and sanitized in a warewashing machine or by immersion in a compartment sink.

NSF and UL Marks
Look for NSF and UL marks on sanitary equipment.

Reprinted by permission of NSF International, Ann Arbor, MI; and Underwriters laboratories, Inc., Northbrook, IL

The Underwriters Laboratories Mark
Look for the Underwriters Laboratories mark for equipment that meets safety standards.

Reprinted by permission of Underwriters laboratories, Inc., Northbrook, IL

- Wooden cutting boards, if allowed by local codes, must be made from non-absorbent hardwood (such as oak or maple), free of seams and cracks, and they must transfer no odor or taste.
- To prevent cross-contamination, use separate cutting boards for raw and cooked foods, and wash and sanitize cutting boards after every use.

CHOOSING AND INSTALLING KITCHEN EQUIPMENT

Well-designed kitchens make the job of keeping food safe easier. Generally, an efficient kitchen design is a more sanitary kitchen design.

- Plan your kitchen layout so that it is easy to clean, minimizes chances for cross-contamination, and minimizes the time foods spend in the temperature danger zone.
- Portable equipment, such as equipment on casters that employees can move, is often easier to clean and clean around than permanently installed equipment.
- Permanently installed equipment must be either mounted on legs at least six inches off the floor or sealed to a masonry base.
- Immobile table-mounted equipment should be mounted on legs that will provide a minimum clearance of four inches between the base of the equipment and the countertop.
- All cracks or seams over $\frac{1}{32}''$ (0.8 mm) must be filled with a nontoxic sealant.

UTILITIES

Utilities used by an establishment include water and plumbing, electricity, gas, lighting, ventilation, sewage, and waste handling. In all cases, there must be enough utilities to meet the cleaning needs of the establishment, and the utility must not contribute to contamination.

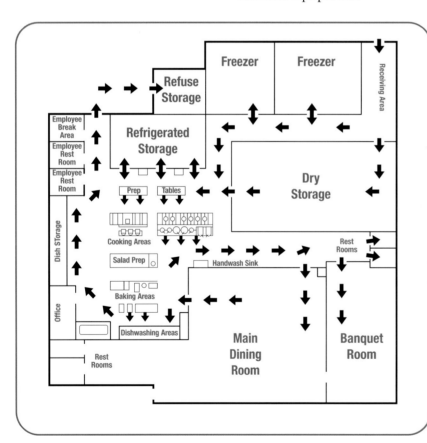

Simplified Foodservice Floor Plan
Arrows indicate normal work flow.

Water Supply

Safe water is vital in an establishment. Unsafe water can carry bacteria, viruses, and parasites. Water that is safe to drink is called potable water. Potable sources of water include the following.

○ Approved public water mains

○ Private water sources that are regularly maintained and tested

○ Bottled drinking water

○ Closed portable water containers filled with potable water

○ On-premise water storage tanks

○ Water transport vehicles that are properly maintained

The use of nonpotable water is extremely limited. If nonpotable water is allowed by local codes, its use is generally limited to the following.

○ Air conditioning

○ Cooling equipment not used for cooling food

○ Fire protection

○ Irrigation (for outdoor grass and plants)

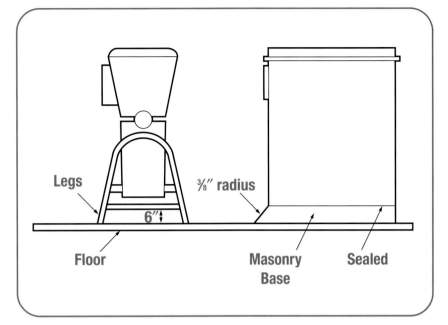

Floor-Mounted Immobile Equipment
Immobile equipment must be mounted on legs at least six inches off the floor or sealed to a masonry base.

Water Emergencies

Occasionally, an emergency may disrupt the water supply. Depending on the nature and severity of the problem, an establishment may wish to remain open. In the case of a foodservice facility in a hospital or shelter, it may need to remain open. If the water supply is disrupted, follow these guidelines.

○ Use bottled or thoroughly boiled water for beverages and ingredients in food recipes.

○ Use commercially prepared ice, if available, or make ice from boiled water.

○ Use boiled water for *essential* cleaning, such as pots and pans. Consider using single-use plates and utensils to minimize washing requirements.

○ Keep a supply of previously boiled warm water available for handwashing.

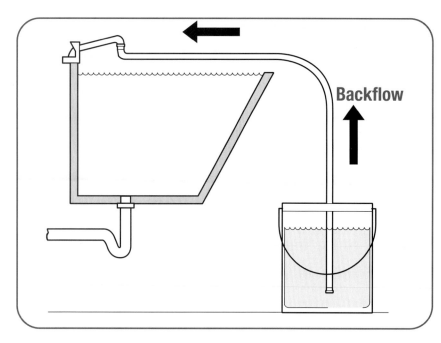

A Common Cross-Connection
A hose connected to a faucet and left submerged in a mop bucket creates a dangerous cross-connection.

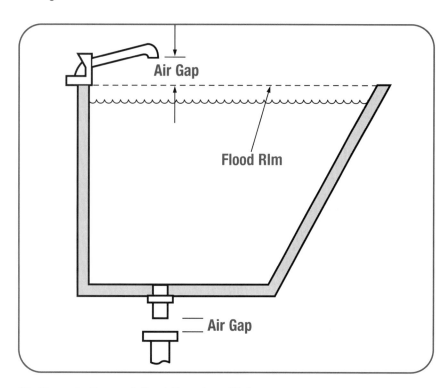

Air Gaps to Prevent Backflow in a Sink
Air gaps between the faucet and the flood rim and between the drainpipe and floor drain of a sink prevent backflow.

Plumbing

Improper plumbing design can cause serious health concerns. Improperly installed or poorly maintained plumbing, which allows the mixing of potable and nonpotable water, has been implicated in outbreaks of typhoid fever, dysentery, Hepatitis A, Norwalk virus, and other gastrointestinal illnesses. Only licensed plumbers should install and maintain plumbing systems in an establishment.

Cross-Connections

The greatest challenge to water safety comes from cross-connections. A cross-connection is a physical link through which contaminants from drains, sewers, or waste water can enter a potable water supply. A faucet located below the flood rim of a sink, or a hose in a mop bucket, are examples of cross-connections.

Cross-connection is dangerous because it allows the possibility of backflow. Backflow is the unwanted reverse flow of contaminants through a cross-connection into a potable water system. It can occur whenever the pressure in the potable water supply drops below the pressure of the contaminated supply.

To prevent cross-connections like this, do not attach a hose to a faucet unless a backflow prevention device, such as a vacuum breaker, is attached. Threaded faucets and connections between two piping systems must have a vacuum breaker or other approved backflow prevention device.

The only completely reliable device to prevent backflow is the air gap. An air gap is an air space that is used to separate a water supply outlet from any potentially contaminated source. A sink may make use of air gaps to prevent backflow. The air space between the faucet and the flood rim of the sink is one air gap. Another may be located between the drainpipe of the sink and the floor drain of the establishment.

Grease Condensation and Leaking Pipes

Grease condensation in pipes is another common problem in plumbing systems. Grease traps are often installed to prevent a buildup of grease from creating a drain blockage. The trap must be cleaned and the grease removed periodically. If this is not done, or is not done properly, an overflow could lead to problems with odor and contamination.

Overhead waste-water pipes or fire safety sprinkler systems can leak and become a source of contamination. Even overhead lines carrying potable water can be a problem, since water can condense on the pipes and drip onto food. All piping should be serviced immediately when leaks occur.

Sewage

Sewage and waste water are contaminated with bacteria, viruses, and parasites. The facility must have adequate drainage to handle all waste water produced. Areas subject to heavy water exposure should have floor drains. Drain pipes carrying waste water or sewage must be clearly identified so they cannot be confused with those carrying potable water. Backups of waste water from sinks or warewashing machines are very serious.

A backup of raw sewage is cause for immediate closure of the establishment, correction of the problem, and thorough cleaning.

Ventilation

Adequate ventilation in food-preparation areas removes odors, gases, and airborne dirt and mold that can cause contamination. Ventilation must be designed so that hoods, fans, guards, and ductwork do not drip onto food or equipment. Hood filters or grease extractors must be tight fitting and easily removable, and should be cleaned on a regular basis. Thorough cleaning of the hood and ductwork should also be done periodically by a professional company. It is the establishment's responsibility to see that the ventilation system meets local regulations.

Solid Waste Management

Waste management is an important issue today in establishments. There are many things that managers may do to improve the waste management problem.

The Environmental Protection Agency (EPA) has recommended three approaches to manage solid waste.

○ Reduce the amount of waste produced. Eliminate unnecessary packaging.

○ Reuse when possible. Reused containers must be cleaned and sanitized. Never reuse chemical containers for food items.

○ Recycle materials. Store recyclables so that they can't contaminate food or equipment or attract pests.

When these approaches are used, the amount of waste can be greatly reduced.

Garbage Disposal

Garbage can be a hazard to an establishment. To control hazards that garbage can pose, consider the following.

○ Garbage should be removed from food-preparation areas as quickly as possible to prevent odors, pests, and possible contamination. Do not carry garbage over or across food or food-preparation areas.

○ Garbage containers must be leakproof, waterproof and pest-proof, and have tight-fitting lids. They should typically be made of galvanized metal or an approved plastic.

○ Plastic or wet-strength paper liners may be used in garbage containers to make garbage removal and cleaning easier.

○ Garbage containers should be cleaned frequently and thoroughly, both inside and out. This will help keep odors and pests to a minimum. Drain plugs in outdoor containers should be kept in place, except during cleaning. Areas used for cleaning garbage containers should not be located near food-preparation or storage areas.

CLEANING AND SANITIZING

It is important to understand the difference between cleaning and sanitizing. Cleaning is the process of removing food and other types of soil from a surface such as a countertop or plate. Sanitizing is the process of reducing the number of microorganisms on that surface to safe levels. To be effective, cleaning and sanitizing must be a two-step process. Surfaces must first be cleaned and rinsed before being sanitized.

Everything in your operation must be kept clean; however, any surface that comes in contact with food must be cleaned and sanitized. All food-contact surfaces must be washed, rinsed, and sanitized:

○ After each use.

○ When you begin working with another type of food.

○ Any time you are interrupted during a task and the tools or items you have been working with may have been contaminated.

○ At four-hour intervals if the items are in constant use.

Cleaning

Cleaning agents are chemical compounds which remove food, soil, rust stains, minerals, or other deposits. Cleaning agents are divided into four categories:

○ **Detergents:** All detergents contain surfactants that reduce surface tension between the soil and the detergent, so the detergent can penetrate and soften the soil.

○ **Solvent cleaners:** Often called degreasers, solvent cleaners are alkaline detergents that contain a grease-dissolving agent. These cleaners work well in areas where grease has been burned on.

○ **Acid cleaners:** Acid cleaners are used on mineral deposits and other soils that alkaline cleaners can't remove. These cleaners are often used to remove scale in warewashing machines and steam tables.

○ **Abrasive cleaners:** These cleaners contain a scouring agent that helps scrub off hard-to-remove soils. These cleaners are used on floors or to remove baked on or burned on foods in pots and pans.

Check with suppliers to find out which compounds are suitable for your needs. Cleaning agents must be stable, non-corrosive, and safe for employees to use. They can be ineffective and even dangerous if misused. Several factors affect the cleaning process. The table on the next page lists these factors and provides a brief explanation of each.

Factors That Affect the Cleaning Process	
Factor	**Affect on Cleaning Process**
Type of Soil	Certain types of soil require special cleaning methods
Condition of Soil	The condition of the soil or stain affects how easily it can be removed. Dried or baked-on stains will be more difficult to remove than soft, fresh stains
	Cleaning is more difficult in hard water because minerals react with the detergent, decreasing its effectiveness
Water Hardness	Hard water can cause scale or lime deposits to build up on equipment, requiring the use of lime-removal cleaners
Water Temperature	In general, the higher the water temperature, the better a detergent will dissolve, and the more effective it will be in loosening dirt
Cleaning Agent and Surface Being Cleaned	Different surfaces require different cleaning agents. Some cleaners work well in one situation, but might not work well or may even damage equipment when used in another
Agitation or Pressure	Scouring or scrubbing a surface helps remove the outer layer of soil, allowing a cleaning agent to penetrate deeper
Length of Treatment	The longer soil on a surface is exposed to a cleaning agent, the easier it is to remove

Sanitizing

There are two methods that can be used to sanitize surfaces, heat sanitizing and chemical sanitizing. Which you use depends on the application.

Heat Sanitizing

The higher the heat, the shorter the time required to kill microorganisms. The most common way to heat sanitize tableware, utensils, or equipment is to immerse or spray the items with hot water. Use a thermometer to check water temperature when heat sanitizing by immersion. Another way to check the water temperature of a machine is to attach temperature-sensitive labels and tapes or a high-temperature probe to the items being cleaned and sanitized.

Chemical Sanitizing

Chemical sanitizing is done in two ways: either by immersing a clean object in a specific concentration of sanitizing solution for a required period of time; or by rinsing, swabbing, or spraying the object with a specific concentration of sanitizing solution.

The three most common types of sanitizer used in the restaurant and foodservice industry are chlorine, iodine, and quaternary ammonium compounds (quats). There are advantages and disadvantages to using each type. The table below lists these advantages and disadvantages.

Scented or oxygen bleaches are not acceptable as sanitizers for food-contact surfaces. Household bleaches are acceptable only if the labels indicate they are EPA registered.

Advantages and Disadvantages of Different Sanitizers		
Types	**Advantages**	**Disadvantages**
Chlorine	Most commonly used sanitizer Kills a wide range of microorganisms Leaves no film on surfaces Least expensive Effective in hard water	Less effective in pH ranges outside 6 to 7.5 Dirt quickly inactivates these solutions Corrosive to some metals such as stainless steel and aluminum when used improperly Adversely affected by temperatures above 115°F (46°C)
Iodine	Effective at low concentrations Not as quickly inactivated by dirt as chlorine Color indicates presence	Less effective than chlorine Less effective at pH levels above 5.0 Becomes corrosive to some metals at temperatures above 120°F (49°C) More expensive than chlorine May stain surfaces
Quaternary ammonium compounds (quats)	Not as quickly inactivated by dirt as chlorine Remains active for a short period of time after it has dried Noncorrosive Nonirritating to skin Works in most temperature and pH ranges	Leaves a film on surfaces Does not kill certain types of microorganisms Hard water reduces effectiveness

Factors Influencing the Effectiveness of Sanitizers

Different factors influence the effectiveness of chemical sanitizers. The most critical include contact time, selectivity, temperature, and concentration. The table on the next page summarizes these factors.

A test kit designed for the specific sanitizer should be used to test its concentration in the sanitizing solution. Test kits are usually available from the manufacturer or a restaurant supplier. The picture to the right shows one in use. A sanitizing solution must be changed when it is visibly dirty, or when the concentration of the sanitizer has dropped below the amount required.

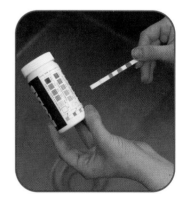

Sanitizer Test Kit
Use the kit to test the concentration of a sanitizing solution.

General Guidelines for Chemical Sanitizers			
	Chlorine	**Iodine**	**Quaternary Ammonium**
Minimum Concentration			
○ For Immersion	50 parts per million (PPM)	12.5-25.0 PPM	220 PPM
○ For Spray Cleaning	50 PPM	12.5-25.0 PPM	220 PPM
Temperature of Solution	Above 75°F (24°C) Below 115°F (46°C)	75°F (29°C) Iodine will leave solution at 120°F (49°C)	Above 75°F (24°C)
Contact Time			
○ For Immersion	7 seconds	30 seconds	30 seconds–read label; some products require longer contact time
○ For Spray Cleaning	Follow manufacturer's directions	Follow manufacturer's directions	
pH (detergent residue raises pH, so rinse completely)	Must be below 8.0	Must be below 5.0	Most effective at 7.0, but varies with compound
Corrosiveness	Corrosive to some substances	Noncorrosive	Noncorrosive
Reaction to Organic Contaminants in Water	Quickly inactivated	Made less effective	Not easily affected
Reaction to Hard Water	Not affected	Not affected	Some compounds inactivated– read label; hardness over 500 PPM is undesirable
Indication of Proper Strength	Test kit required	Amber color indicates presence. Use test kit to determine concentration	Test kit required; follow directions closely

MACHINE WAREWASHING

Most tableware, utensils, and even pots and pans can be cleaned and sanitized in warewashing machines. Most warewashing machines sanitize by using either hot water or a chemical-sanitizing solution.

High-Temperature Machines

○ These machines rely on hot water to clean and sanitize.

○ The temperature of the final sanitizing rinse must be at least 180°F (82°C). For stationary-rack single-temperature machines, the temperature of the final sanitizing rinse must be at least 165°F (74°C).

○ Make sure your warewasher has a built-in thermometer to measure the temperature of water at the manifold, where it sprays into the tank.

Chemical-Sanitizing Machines

○ These machines use chemicals to sanitize. Warewashing machines that use chemical sanitizing often wash at much lower temperatures, but not less than 120°F (49°C).

○ Rinse-water temperature in these machines should be between 75°F and 120°F (24°C to 49°C) for the sanitizer to be effective.

Warewashing Machine Operation Guidelines

All warewashing machines should be operated according to manufacturers' instructions. No matter what type of machine you use, there are some general procedures to follow to clean and sanitize tableware, utensils, and related items.

○ **Check the machine for cleanliness and clean it as often as needed, at least once a day.** Fill tanks with clean water. Clear detergent trays and spray nozzles of food and foreign objects. Use an acid cleaner on the machine at least once a week to remove mineral deposits caused by hard water.

○ **Make sure detergent and sanitizer dispensers are properly loaded.**

○ **Scrape, rinse, or soak items before washing.** Pre-soak items with dried-on food.

○ **Load warewasher racks correctly and use racks designed for the items being washed.** Make sure that all surfaces are exposed to the spray action. Never overload racks.

○ **Check temperatures and pressure.** Follow manufacturers' recommendations.

○ **Check each rack for soiled items as it comes out of the machine.** Run dirty items through again until they are clean. Most items will need only one pass if you use proper equipment and procedures.

○ **Air dry all items.** Towels may contaminate items.

○ **Keep your warewashing machine in good repair.** Some common problems can be solved easily by changing procedures or chemicals.

MANUAL WAREWASHING

Establishments that do not have a warewashing machine may use a three-compartment sink to wash items (some local regulatory agencies allow the use of two-compartment sinks; others require four-compartment sinks). These sinks may also be used to wash larger items as well. A properly set up warewashing station will include:

○ An area for scraping or rinsing food into garbage containers.

○ Drainboards to hold both soiled and clean items.

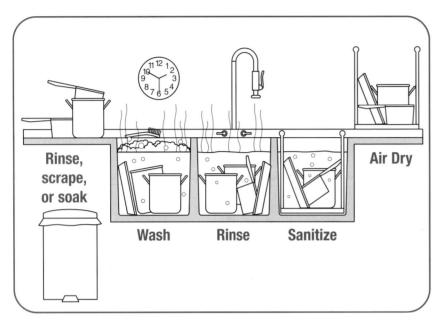

Three-Compartment Sink for Manual Washing, Rinsing, and Sanitizing

○ A thermometer in each sink to measure water temperature.

○ A clock with a second hand. This will allow employees to time how long items have been immersed in the sanitizing sink.

Before washing items, clean and sanitize each sink and all work surfaces. Follow these steps when washing and sanitizing all tableware, utensils, and equipment.

Step 1: Rinse, scrape, or soak all items before washing.

Step 2: Wash items in the first sink in a detergent solution. Water temperature should be at least 110°F (43°C). Use a brush, cloth, or nylon scrubber to loosen remaining soil. Replace detergent solution when suds are gone or water is dirty.

Step 3: Immerse or spray-rinse items in second sink. Water temperature should be at least 110°F (43°C). Remove all traces of food and detergent. If using immersion method, replace water when it becomes cloudy or dirty.

Step 4: Immerse items in third sink in hot water or a chemical-sanitizing solution. If hot water immersion is used, the water temperature must be at least 171°F (77°C). Some codes require a temperature of 180°F (82°C). Items must be immersed for thirty seconds. If chemical sanitizing is used, the sanitizer must be mixed at the proper concentration. (Check at regular intervals with a test kit.) Water must be the correct temperature for the sanitizer used.

Step 5: Air dry all items on a drainboard.

CLEANING AND SANITIZING EQUIPMENT

Equipment must be kept clean. All food-contact surfaces must be sanitized as well. Teach all employees how to properly clean and sanitize each type of equipment.

Stationary Equipment

Equipment manufacturers will usually provide cleaning instructions. Food-contact surfaces may require a different cleaning or sanitizing solution than non-food-contact surfaces. In general, follow these steps.

○ Turn off and unplug equipment before cleaning. Refrigerators and freezers should be turned off but can be left plugged in.

○ Remove food and soil from under and around the equipment.

○ Remove detachable parts and manually wash, rinse, and sanitize them, or run them through a warewasher if permitted. Allow them to air dry. When washing sharp parts such as slicer blades, turn the blades away from yourself and wipe away from sharp edges.

○ Wash and rinse fixed food-contact surfaces, then wipe with chemical-sanitizing solution.

○ Keep cloths used for food-contact and non-food-contact surfaces in separate, properly marked containers of sanitizing solution.

○ Air dry all parts, then reassemble according to directions. Tighten all parts and guards. Test equipment at recommended settings, then turn off.

○ Re-sanitize food-contact surfaces that were handled when putting the units back together, by wiping with a cloth that has been submerged in sanitizing solution.

In some cases, you may be able to spray-clean fixed equipment. Check with the manufacturer first. If allowed, spray each part with solution in the right concentration for two or three minutes.

NON-FOOD STORAGE

Once tableware, utensils, and equipment are clean and sanitary, store them so they stay that way. Storage areas for cleaning supplies should be out of the way of kitchen traffic and potential cross-contamination. Keep storage areas clean and sanitary, too.

Tableware and Equipment

○ Store tableware and utensils at least six inches off the floor. Keep them covered or otherwise protected from dirt and condensation.

○ Clean and sanitize drawers and shelves before clean items are stored.

○ Clean and sanitize trays and carts used to carry clean tableware and utensils to and from the storage area. Do this daily or more often if they become soiled.

○ Store glasses and cups upside down. Store flatware and utensils with handles up or out so employees can pick them up by the handles.

○ Keep food-contact surfaces of clean-in-place equipment covered until ready to use.

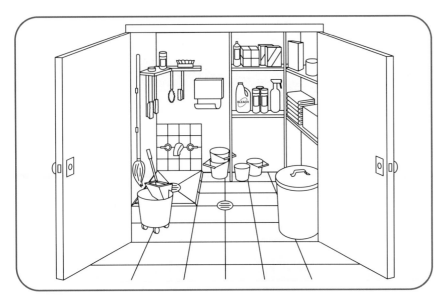

Storage Area for Cleaning Tools and Supplies

Cleaning Tools and Supplies

Cleaning tools and supplies should be cleaned and sanitized before being put away. Tools and chemicals should be stored in a locked area away from food and prep areas. The area should be well lighted so employees can identify chemicals easily.

USING HAZARDOUS MATERIALS

Chemicals are both useful and necessary to keep an establishment clean, sanitary, and free of pests. Used properly, they may pose little threat to an employee's safety. Used improperly, they may become a health hazard that can cause injury.

Because of the potential danger of chemicals used in the workplace, the Occupational Safety and Health Administration (OSHA) requires employers to comply with their Hazard Communication Standard (HCS). This standard, also known as Right-to-Know or HAZCOM, requires employers to tell their employees about chemical hazards they may be exposed to at the establishment. It also requires employers to train employees how to safely use the chemicals they work with. Employers must comply with OSHA's HCS standard by developing a hazard communication program for their establishment.

A hazard communication program must include the following components:

○ An inventory of the hazardous chemicals used at the establishment

○ Chemical labeling procedures

○ Material Safety Data Sheets (MSDS)

○ Employee training

○ A written plan addressing hazard communication standards

Material Safety Data Sheets (MSDS)

OSHA requires that chemical manufacturers and suppliers provide a Material Safety Data Sheet (MSDS) for each hazardous chemical at your establishment. These sheets are a part of employees' right to know about the hazardous chemicals they work with and must be kept in an accessible location. MSDS sheets contain the following information about the chemical.

○ Information about safe use and handling

○ Physical, health, fire, and reactivity hazards

○ Precautions

○ Appropriate personal protective equipment (PPE) to wear when using the chemical.

○ First-aid information and steps to take in an emergency

○ Manufacturer's name, address, and phone number

○ Date the MSDS was prepared

○ Hazardous ingredients and identity information

IMPLEMENTING A CLEANING PROGRAM

A clean and sanitary environment is a prerequisite to an effective HACCP-based food-safety program. A cleaning program will give you a system to organize all your cleaning and sanitizing jobs. As with HACCP, an effective cleaning program requires commitment from management and involvement from employees. Here are some basic steps to design and implement a cleaning program.

Create a Master Cleaning Schedule

Walk through every area of the facility and write down all surfaces, tools, and equipment that need to be cleaned. Get input from employees during your walk-through and develop a master cleaning schedule. The schedule should include the following.

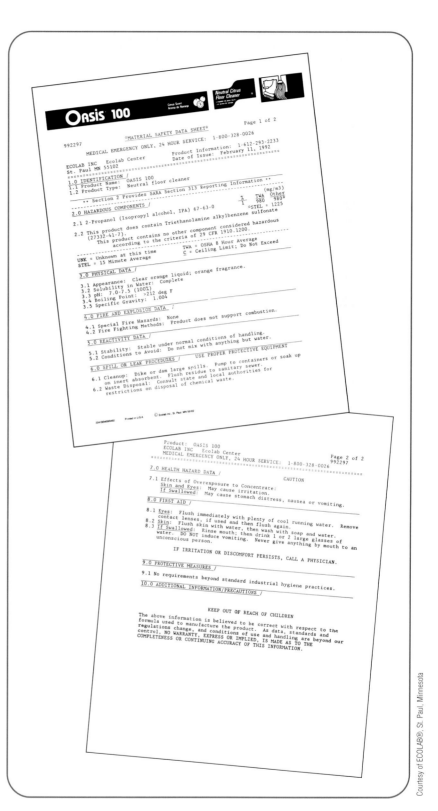

A Sample Material Safety Data Sheet

○ **What should be cleaned.** Arrange the schedule in a logical way so nothing is left out. List all cleaning jobs in one area, or list jobs in the order they should be performed.

○ **Who should clean it.** Assign each task to a specific individual.

○ **When it should be cleaned.** Employees responsible for their own areas should clean as they go and clean and sanitize at the end of their shifts. Schedule major cleaning when food won't be contaminated and service won't be interrupted—that usually means after closing. Schedule work shifts to allow enough time for cleaning.

○ **How it should be cleaned.** Provide clearly written procedures for cleaning. Lead employees through the process step by step. Always follow manufacturers' instructions when cleaning equipment. Specify cleaning tools and chemicals by name.

Monitor the Program

○ Supervise daily cleaning routines.

○ Monitor completion of all cleaning tasks daily against the master cleaning schedule.

○ Review and change the master schedule every time there is a change in menu, procedures, or equipment.

○ Request employee input on the program during staff meetings.

○ Conduct spot inspections.

INTEGRATED PEST MANAGEMENT

Pests such as insects and rodents can pose serious problems for establishments. Not only are they unsightly to customers, they also damage food, supplies, and facilities. The greatest danger from pests comes from their ability to spread diseases, including foodborne illnesses.

Once pests have infested a facility, it can be very difficult to eliminate them. Developing and implementing an integrated pest management (IPM) program will help prevent pests from infesting your establishment. It will require an ongoing and continuous program to get rid of any pests that do invade the premises.

For your IPM program to be successful, you must work closely with a licensed pest control operator (PCO). These professionals use safe, up-to-date methods to effectively prevent and control pests. Prevention is critical in pest control. If you wait until there is evidence of pests in your establishment, you may already have a major infestation.

THE IPM PROGRAM

There are three basic rules of an IPM program.

○ Deny pests access to the facility.

○ Deny pests food, water, and a hiding or nesting place.

○ Work with a licensed and registered PCO to eliminate pests that do enter.

Denying Pests Access

Pests can enter an establishment in one of two ways. They are either brought inside with deliveries, or they enter through openings in the building itself. To prevent pests from entering your establishment, pay particular attention to the following areas.

Deliveries:

○ Use reputable suppliers.

○ Check all deliveries before they enter your establishment.

○ Refuse shipment that have signs of a pest infestation, for example, egg cases and body parts (legs, wings, etc.).

Doors, Windows, and Vents:

○ Screen all windows and vents with at least sixteen mesh per square inch screening.

○ Install self-closing devices or door sweeps on all doors.

○ Install air curtains (also called air doors or fly fans) that blow a steady stream of air, creating an air shield around doors that are left open.

○ Keep all exterior openings closed tightly. Check them for proper fit as part of a regular cleaning schedule.

Pipes

Mice, rats, and insects such as cockroaches use pipes as highways through a facility.

○ Use concrete to fill holes or sheet metal to cover openings around pipes.

○ Install screens over ventilation pipes and ducts on the roof.

Floors and Walls

Rodents often burrow into buildings through decaying masonry or cracks in building foundations. They move through floors and walls the same way.

○ Seal all cracks in floors and walls. Use a permanent sealant recommended by your PCO, local health department, or building contractor.

○ Properly seal spaces or cracks where fixed equipment is fitted to the floor.

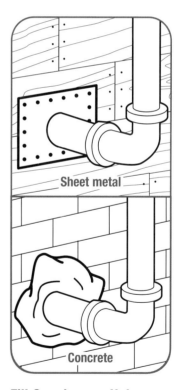

Sheet metal

Concrete

Fill Openings or Holes Around Pipes with Sheet Metal or Concrete to Deny Entry to Pests

○ Paint a white stripe around the edge of your storeroom that extends six inches from the walls. This stripe will remind employees to store supplies six inches from the walls. It will also help you monitor the area for signs of infestation, since hairs, tracks, and droppings will show up better against the white stripe.

○ Cover floor drains with hinged grates.

Deny Food and Shelter

Pests are usually attracted to damp, dark, dirty places. A clean and sanitary establishment offers them little in the way of food and shelter. The stray pest that might get in cannot thrive or multiply in a clean kitchen. Start with these preventive measures to deny pests food and shelter:

○ Dispose of garbage quickly and correctly. Keep garbage containers clean, in good condition, and tightly covered in all areas (indoor and outdoor). Clean up spills around garbage containers immediately. Wash, rinse, and sanitize containers regularly.

○ Store recyclables in clean, pest-proof containers away from your building.

○ Properly store all food and supplies as quickly as possible.

- Keep all food and supplies at least six inches off the floor and six inches away from the walls.

- When possible, keep humidity lower than 50 percent. Low humidity helps prevent roach eggs from hatching.

- Refrigerate foods such as powdered milk, cocoa, and nuts after opening. These foods attract insects, but most insects become inactive at temperatures below 41°F (5°C).

- Use FIFO, so pests don't have time to settle into these products and breed.

○ Clean and sanitize the facility thoroughly. Careful cleaning eliminates the food supply, destroys insect eggs, and reduces the number of places pests can safely take shelter.

IDENTIFYING PESTS

Pests may still get into your establishment even if you take careful preventive measures. Learn how to spot signs of pests and identify what kind they are. If possible, record the time, date, and location of any pest sighting (or evidence of pests) and report this to your PCO. Early detection gives your PCO a chance to start treatment as soon as possible.

Cockroaches

Roaches often carry disease-causing microorganisms such as *Salmonella,* fungi, parasite eggs, and viruses.

They generally feed in the dark. If you see a cockroach in daylight, you may have a major infestation. Only the weakest roaches come out in daylight. If you suspect you have a roach problem, check for these signs.

- A strong oily odor
- Droppings (feces), which look like grains of black pepper
- Egg cases, which are capsule shaped; brown, dark red, or black; and may appear leathery, smooth, or shiny

Glue Traps

You may have problems with more than one type of roach. Glue traps should be used to monitor your establishment and find out what type of roaches might be present. These open-ended cardboard containers have sticky glue on the bottom that traps roaches. When using glue traps:

- Place the traps in places that roaches can typically be found.
- If possible, place traps on the floor in the corner where two walls meet.
- Check the traps after twenty-four hours and show them to your PCO. The type and stage (nymph or adult) of roaches will determine what type of treatment is needed.

A Roach Glue Trap
Glue traps can be placed in areas where roaches are typically found to monitor their presence.

Courtesy of the National Pest Control Association

Rodents

Rodents are a serious health hazard. They eat and ruin food, damage property, and can spread disease.

A building may be infested with both rats and mice at the same time. Look for these signs.

- **Droppings.** Fresh droppings are shiny and black. Older droppings are gray.
- **Signs of gnawing.** Rats and mice gnaw to reach food.
- **Tracks.** Check dusty surfaces by shining a light across them at a low angle.
- **Nesting materials.** Mice use scraps of paper, cloth, hair, and other soft materials to build nests.
- **Holes.** Rats nest in burrows, usually in dirt, rock piles, or along foundations.

CONTROL MEASURES

PCOs can use a variety of pest-control methods that are environmentally sound and safe for establishments. They are trained to know which techniques will work best to control different types of pests in your area.

Controlling Insects

There are several ways to control insects, depending on the type of insect and degree of infestation.

○ **Repellents.** Repellents are liquids, powders, or mists that keep insects away from an area, but do not kill them. Repellents are often used in hard-to-reach places, such as behind wall-boards and plaster.

○ **Sprays.** Chemical pesticide sprays are used to control roaches, flies, and ants. Because you can buy them in any supermarket, they are easy to misuse and abuse. Let your PCO, not your employees, use spray to control insects. Letting employees apply chemicals is risky, and improperly applied pesticides may be ineffective. Prepare the area to be sprayed by removing all foods and food-contact utensils. Cover equipment and food-contact surfaces that can't be moved. Wash, rinse, and sanitize food-contact surfaces after the area has been sprayed.

  ● Residual sprays leave a film of insecticide that insects absorb as they crawl across it. They are used in cracks and crevices like those along baseboards. Sprays can be liquid or a dust, such as boric acid.

  ● Contact sprays kill insects on contact. They are usually used on groups of insects, such as clusters of roaches or a nest of ants.

○ **Bait.** Chemical bait is sometimes used to control roaches or ants. The bait contains an attractant. When insects eat it, the chemical kills them.

○ **Traps.** Traps are generally used for flying insects such as flies and mosquitoes. There are several types. Wasp and hornet traps are designed to hold nectar, which attracts them. Once inside they cannot escape. Most fly traps use a light source, usually UV light, to attract insects to the trap.

The placement of traps is important. Never place them above or near food-preparation areas or food-contact surfaces.

Controlling Rodents

Rats and mice tend to use the same routes in your facility. Your PCO will choose the best method to eliminate these pests.

○ **Traps.** Traps are one safe, effective way to kill rats and mice. If the infestation is large, however, traps will take time. Spring traps use food, such as peanut butter, as bait. Food should be kept fresh. Set traps near or

Glue Board

Multi-use traps

Mouse and Rat traps

Methods for Controlling Rodents

Several devices can be used to control rodents, including glue boards and traps.

Courtesy of The National Pest Control Association.

in rodent runways. Check traps often, and remove dead rodents carefully. If a trapped rodent is still alive, have your PCO remove it.

○ **Glue boards.** Glue boards work for killing mice. When these devices are placed in runways, mice get stuck to the board and die in several hours from exhaustion or lack of water or air. Check the boards often. Throw away any with trapped mice. Glue boards are not effective for controlling rats. Rats are strong enough to escape the boards.

○ **Bait.** Chemical bait should be used only by a PCO. Chemical bait should be used outdoors where it cannot contaminate food or food-contact surfaces. It is usually placed in special covered, locked containers near rodent runways and possible entry points. Your PCO may change the baits and their locations often until they begin to work. Rats can easily detect chemical bait and often avoid it.

USING AND STORING PESTICIDES

Rely on your PCO to decide if and when pesticides should be used in your establishment. PCOs are trained to determine the best pesticide for each pest, and how and where to apply it.

Your PCO should store and dispose of all pesticides used in your facility. If you store pesticides, follow these guidelines.

○ **Keep pesticides in their original containers.**

○ **Store pesticides in locked cabinets away from food-storage and food-preparation areas.** Store them separately from cleaning supplies.

○ **Store aerosol or pressurized spray cans in a cool place.** Exposure to temperatures higher than 120°F (49°C) could cause them to explode.

○ **Check local regulations before disposing of pesticides.** Many are considered hazardous waste.

○ **Dispose of empty containers according to manufacturers' directions and local regulations.**

○ **Keep a copy of the corresponding MSDS on the premises.**

Only PCOs Should Apply Pesticides at Your Establishment

Courtesy of The National Pest Control Association.

SUMMARY

An establishment that is difficult to clean will not be cleaned well. Sanitation efforts will be more effective if the establishment is designed and equipped with easy cleaning in mind. Equipment must meet the sanitation standards set by NSF International or the equivalent and should be placed so that all areas around and under it can be cleaned effectively.

Cleaning is the process of removing visible soil with a cleaning agent and agitation. Sanitizing is the process of reducing the number of harmful microorganisms to a safe level. You must clean and rinse a surface before it can be sanitized effectively. Surfaces can be sanitized with hot water at least 171°F (77°C) or with a chemical-sanitizing solution. Food-contact surfaces must be cleaned and sanitized after every use, every four hours if in continuous use, or at least once every day.

Follow manufacturers' instructions when using warewashing machines. Check temperatures and pressure of wash and rinse cycles daily. Manual warewashing may be done in a compartment sink, or, if the items are immovable, by cleaning and then spraying them with a sanitizing solution. Items that are cleaned in a three-compartment sink should be pre-soaked or scraped clean, washed in a detergent solution, rinsed in clear water, and sanitized in either hot water (at 171°F or 77°C) or immersed in a sanitizing solution for a predetermined amount of time. All items should then be air dried.

Cleaning tools and supplies should be stored in a well-lit locked room separate from food storage and preparation areas. Chemicals should be clearly labeled, and MSDS should be on hand for each chemical. Create a master cleaning schedule that lists all cleaning tasks and when and how they are to be performed. Assign responsibility for each task by job title. Monitor the cleaning program to keep it effective.

Pests are clearly a menace to establishments because they can carry and spread a variety of diseases. Work with a pest control operator (PCO) to develop and implement an integrated pest management (IPM) program in your establishment. IPM utilizes a combination of preventive and control measures to eliminate pests and keep them from infesting your facility. Preventive measures focus on two areas: denying pests access to the facility and eliminating sources of shelter and food.

If pests are detected, control measures may be necessary. Both chemical and nonchemical methods may be used to control pests. Most chemical pesticides are toxic to humans and should be applied only by a licensed or certified pest control operator.

ACTIVITY

Crossword Puzzle

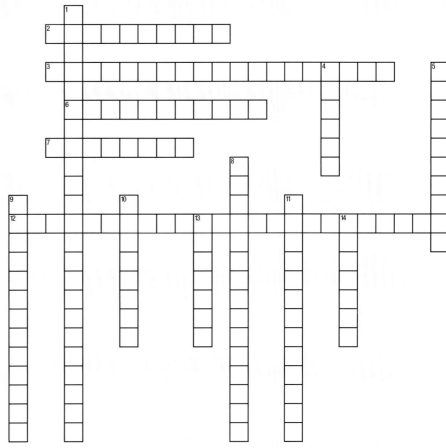

Across:

2. The process of reducing the number of microorganisms on a surface to safe levels.

3. A licensed or certified technician who implements and monitors pest-control programs for companies that contract for services.

6. The situation when pests over-run or inhabit an establishment in large numbers.

7. The process of removing food and other types of soil from a surface such as a countertop or plate.

12. A system of preventive measures and control measures to prevent pests from entering your facility, and to eliminate existing pest infestations.

Down:

1. A sheet that lists the chemical and common name of the material, potential physical and health hazards, and directions for safe handling and use.

4. An air space separating a water supply outlet from any potentially contaminated source. It is the only completely reliable method for preventing backflow.

5. Solid or semi-solid material resulting from food preparation or service that is unwanted or hazardous, some of which may be recyclable.

8. A physical link through which contaminants from drains, sewers, or waste water can enter a potable water supply.

9. An article designed to be used once and thrown away, such as paper towels, plastic gloves, paper plates, plastic eating utensils, and some condiment packets.

10. The unwanted reverse flow of contaminants through a cross-connection into a potable water system.

11. A mechanical device for preventing backflow between two water systems.

13. Water that is safe to drink or use as an ingredient in food.

14. Wet waste material, usually containing food, that cannot be recycled. It attracts pests and has the potential to contaminate food items, equipment, and utensils.

ACTIVITY

Word Find

Find the terms that go with the clues below.

Clues

1. Water that is safe to drink or use as an ingredient in food.

2. Wet waste material, usually containing food, that cannot be recycled. It attracts pests and has the potential to contaminate food items, equipment, and utensils.

3. Solid or semi-solid material resulting from food preparation or service that is unwanted or hazardous, some of which may be recyclable.

4. The process of removing food and other types of soil from a surface such as a countertop or plate.

5. The process of reducing the number of microorganisms on a surface to safe levels.

6. A system of preventive measures and control measures to prevent pests from entering your facility, and to eliminate existing pest infestations.

7. An air space separating a water supply outlet from any potentially contaminated source. It is the only completely reliable method for preventing backflow.

8. A physical link through which contaminants from drains, sewers, or waste water can enter a potable water supply.

9. The unwanted reverse flow of contaminants through a cross-connection into a potable water system.

10. A mechanical device for preventing backflow between two water systems.

11. The situation when pests overrun or inhabit an establishment in large numbers.

12. A licensed or certified technician who implements and monitors pest-control programs for companies that contract for services.

13. A sheet that lists the chemical and common name of the material, potential physical and health hazards, and directions for safe handling and use.

14. An article designed to be used once and thrown away, such as paper towels, plastic gloves, paper plates, plastic eating utensils, and some condiment packets.

```
H S G A D N T H O B R B N U Q G P J A H G Y N I
Y M A A H Y S X X X P O I S V S X D D C X O N S
X S A N R H J P X T W E Z F F A I M W J I T A I
L O T T I B F I F C N V T X U C D V M T E U U N
B A E N E T A H E I C G T M G M P G A G N B H G
E P F U E R I G O X X T E G S B D T R Z M E F L
N X D O S M I Z E N S R Q N C G S A X B J O K E
O B Y Z Z B L A I N S F C D N E T Y H V T B Y U
I E I R X D J E L N K Z G I F E S T V X M P P S
T Q I Y N K Q G S S G D N N D Z Q U H Z O O J E
C R K M F R B V J C A A I P F S C J O J G Q M I
E S I G I V V H S U E F E A Q W R P B W U R N T
N R F Q F D A B Q L S S E O I O U F T Z J C M E
N Y Z B U J I H C O T G L T T R T R F K X K P M
O V B G B L K N L M J W S G Y Z G H L K Q G S G
C X L C P I E I A N Z G L M S D W A Y C X F F K
S Y Y B B L D N N W I F T L I Q A O P K C L W D
S K A F B W A V D D W X E H N J R T L L A S X M
O V R A A G L B J J O S L R S X M Z A F D D I J
R K T S E Z K X Q J J Z X D S K O W E S K Y C I
C O T M H M T J H C Q N D D K F Q F Z H H C O O
P E E V B B E U D S P T U N N J R I F R Y E A B
V N N N B G Q D J R E K A E R B M U U C A V E B
T P E S T C O N T R O L O P E R A T O R R P J T
```

ACTIVITY

What's Wrong with This Picture?

There are several food-safety problems illustrated in the picture below. How many can you find?

ACTIVITY

Water Lines Burst...Story at 11 p.m.!

It is 6:00 p.m. on the eve of the busiest day for your establishment. You have just been notified that the water supply lines on your block have burst due to a construction accident and service will be interrupted for twenty-four hours. This will leave you without potable water until 6:00 p.m. the next day. Since tomorrow is the busiest day of the year for you, you must continue to conduct business as usual. Outline the things you need to consider to continue to operate without a potable water supply.

ACTIVITY

Can You Identify the Sanitizer?

Directions: For each of the characteristics listed below, identify the sanitizer that best matches the characteristic.

Example: Effective in hard water   _____*Chlorine*_____

Characteristic: **Sanitizer:**

1. Most commonly used sanitizer in restaurant and establishments.   _____

2. This sanitizer may stain equipment as well as clothing.   _____

3. In the presence of dirt, this sanitizer is quickly inactivated.   _____

4. If money is an object, this is your best bet.   _____

5. Keeps working after it appears that it has gone away. (Only for a short time!)   _____

6. If there is no color, this sanitizer isn't around.   _____

7. If the surface is visibly different after using this sanitizer, then this sanitizer has been used.   _____

8. When this sanitizer is not used properly, it can be a nightmare on a metal countertop or even a Coke can!   _____

9. This sanitizer is not as aggressive as the others; it won't kill everything.   _____

10. When this sanitizer gets hot—120°F (49°C)—it's heavy on metal!   _____

11. You don't have to worry about hurting your hands with this sanitizer.   _____

12. Metals don't fear this sanitizer.   _____

13. A little dab will do you.   _____

14. This sanitizer is more effective at a neutral pH.   _____

15. Less effective and more expensive than chlorine.   _____

ACTIVITY

MSDS Quiz

Directions: Using the MSDS sheet provided by your instructor, answer the following questions.

- What PPE should be worn when using this chemical?
- What should you do if you get this chemical on your skin?
- What is the manufacturer's phone number?
- How might this product react with water?
- Are there fire hazards in dealing with this product?
- What respiratory problems are associated with this product?
- What conditions must be avoided when using this product?
- How should this product be disposed of?
- What should be done if this product is ingested?
- What is a restriction in using this product?

ACTIVITY

How Do We Control the Pest?

A. For the pest-control measures listed below, identify whether the measure is used to control rodents or insects by placing an "R" or an "I" next to the control measure.

Control Measure:

1. Spring trap   _____

2. Locked bait container   _____

3. Nectar trap   _____

4. Liquid repellent   _____

5. Glue board   _____

6. UV trap   _____

B. Not done yet! Write a brief statement identifying the proper placement of each control measure.

1. _____

2. _____

3. _____

4. _____

5. _____

6. _____

ACTIVITY

PCO Coming to Your Establishment...

You and your fellow group members have just inherited a small amount of money, $1.5 million to be exact. Having an entrepreneurial spirit and great knowledge of the restaurant and foodservice industry, you decide to start your own business. All of you have more than ten years' experience in the restaurant business and know the importance of pest prevention and control in an establishment. You decide to purchase and manage a pest control operator (PCO) service. The group agrees that the first order of business is to market this service to the industry under a new name. In the space below, create a magazine or newspaper advertisement that will promote the benefits your business can provide. Keep in mind that an establishment is most concerned with preventing pests from getting into the facility and controlling them once inside. This is your chance to show that your business is better than the competition. Good luck!

A CASE IN POINT

Case Study

Several people became ill shortly after drinking beverages at a local restaurant with a bar lounge. They all complained that their iced drinks had an odd taste. At this time in the lounge, the glassware washing machine had been out of service.

When interviewed, the manager explained that the machine's wash cycle was functional but that the unit could not be used because the large volume of water discharged after each wash load was worsening a recent drain blockage problem. A maintenance worker mentioned that there had been intermittent backups in the plumbing during the previous week and a large pool of water was found under the glassware washing machine. Drain cleaners had been used repeatedly with no change in the blockage.

The beverage ice-making machine shared piping with the glassware washing machine in the bar and the grease trap on the sink in the restaurant. The manager revealed that he had installed the plumbing fixtures himself.

When ice cubes were removed from the beverage ice bin, congealed food grease and food debris were found on them. Grease and debris also covered the bottom of the ice bin.

Why do you think the people became ill? What should the manager do to correct the problem?

NOTES

TRAINING TIPS

Training Tips for the Classroom

1. *Expect the Unexpected: Crisis Management Group Activity*

Objective: *After completing this activity, class participants will be able to identify food-safety risks associated with various operational emergencies and develop action plans to address them.*

Directions: List a number of possible crisis situations that could occur in an establishment related to the following topics in Section 10. These may include the following:

○ **Plumbing:** broken or stopped-up pipes or drains; cross-connections; backup of waste water or raw sewage

○ **Water:** water turned off for eight hours or longer; contamination of city or private water supply

○ **Electric:** a power outage for eight hours or longer

○ **Equipment failure:** a broken booster heater in warewashing machine; freezer or cooler malfunction; water-heater breakdown

○ **Natural disaster:** major flooding, hurricane, earthquake

○ **Waste management:** a strike by sanitation workers

Break the class into groups. Assign each group a specific crisis and give them this set of directions.

○ Determine the specific food-safety risk or risks, if any, involved with the crisis.

○ Create an action plan to address the crisis.

○ Develop a plan to deal with, or prevent a future occurrence of, that specific crisis.

Allow twenty minutes for this activity, then have a spokesperson for each group present their work. Solicit feedback and additional suggestions from the other groups after each spokesperson has finished speaking.

Allow time for students to tell about any personal experiences related to crisis situations such as those discussed. Conclude the activity by asking students if their establishment has adequate crisis management plans and, if not, what steps they will take to implement such plans.

2. Let's Design and Build a Restaurant

Objective: *After completing this activity, class participants will be able to identify food-safety issues that should be considered when designing, constructing, and equipping an establishment.*

Directions: Present an establishment concept (example: full-service Italian) and menu to the class. Provide the class with the proposed establishment's hours of operation and the projected volume of business (number of meals served). Include the possibility of additional business from carryout and catering sales.

Break the class into five groups, and assign each group one of the following topics that were discussed in Section 10.

○ Design and layout

○ Equipment selection

○ Plumbing, water, and electric

○ General construction (floors, walls, ceiling)

○ Heating, venting, air conditioning

Direct the groups to develop an action plan that includes a list of food-safety concerns and issues related to their topic. Allow ten minutes for this activity. Then allow time for members of each group to share ideas with other groups, ask questions about each other's plans, and so on, for another ten minutes.

Ask each group to make a brief presentation on its action plan. Solicit class discussion. Conclude this activity by reminding the class of all the food-safety issues related to these areas, and reiterate how a well-designed and efficiently operated establishment can contribute to a successful food-safety program.

3. The Clean Team Competition

Objective: *After completing this activity, class participants should be able to write step-by-step cleaning and sanitizing procedures for kitchen areas and equipment.*

Directions: Break your class into groups of two to four people each, and ask each group to write out the procedures for cleaning and sanitizing a specific piece of equipment in the kitchen. Pick a piece of equipment that is common to most people in the class.

The groups should list the specific cleaning and sanitizing agents they would use, the specific tools, and the step-by-step procedures for the process, from start to finish.

After five or ten minutes, have the teams describe their cleaning and sanitizing procedures. Compare the different teams' work, and award a prize to the team whose procedures were the clearest, most thorough, and accurate.

4. Help Wanted: PCO!

Objective: *After completing this activity, class participants will be able to identify prevention and control measures for pest infestations.*

Directions: Ask the class for three volunteers. These volunteers are PCOs (pest control operators). The class will assume the role of foodservice managers in charge of hiring a new pest control operator.

While the three PCOs are waiting outside the classroom, ask the class to develop a series of interview questions that focus on testing the PCOs' knowledge of preventing and controlling pest infestations. These may include questions such as the following.

○ How can you determine if an establishment has a cockroach infestation?

○ How would you control a cockroach infestation?

○ How would you prevent a rodent infestation?

After interviewing each of the three candidates separately, have the class evaluate the responses and make a choice. Bring all three PCOs back into the classroom, announce the winner, and let a spokesperson for the class explain why the candidate was chosen.

Training Tips on the Job

1. Develop a "Waste Management Task Force"

Purpose: *To involve staff members in an effort to improve the waste-handling program at your establishment. Note: Involving your staff and allowing them to initiate approved changes in the existing program will create buy-in.*

Directions: Solicit a team of employees to perform the following research and development assignment.

○ Assess the waste-handling methods currently being used in the establishment. Consider all aspects: number, type, and conditions of garbage cans; location and condition of outside garbage bins; other waste-handling equipment; the recycling program, and so on.

○ Research and compare waste-handling methods in other operations similar to yours.

○ Read up on local health department codes and regulations regarding waste handling.

○ Develop recommendations for improved waste-handling techniques and systems, along with a list of expected benefits from the proposed improvements.

Give the team some clear direction and a deadline, and keep in touch with its progress along the way. When the report and recommendations are complete, schedule a formal meeting with the team to receive its proposals.

Authorize the team to initiate any approved changes. Provide funding if necessary. Monitor the results, offer suggestions, and solicit feedback from the team. Acknowledge the team's effort with an appropriate reward.

2. Creating a Cleaning Program

Purpose: *To enlist the help of employees from different areas of the establishment to develop a master cleaning schedule.*

Directions: Enlist a "clean team" of employees from different positions in your establishment. Work closely with the team to design and implement an effective, workable, comprehensive master cleaning schedule. Assign team members to work with you on the following tasks.

○ Identifying cleaning needs. Walk through the establishment with team members and identify tools, equipment, and surfaces that need to be cleaned.

○ Create a master cleaning schedule. Identify what should be cleaned, how it should be cleaned, when it should be cleaned, and by whom.

○ Select cleaning tools and supplies.

Once the master cleaning schedule has been created, schedule a kick-off meeting with all staff to introduce the program. Empower "clean team" members to train their coworkers in the new cleaning and sanitizing procedures in their department. Give team members an appropriate reward for their hard work. Schedule periodic meetings with the "clean team" to monitor the progress of the cleaning program. Modify the program if necessary.

3. "How To HAZCOM"

Purpose: *To enlist the help of employees to assess the effectiveness of the establishment's Hazard Communication Program and to modify the program if deficiencies are found.*

Directions: Enlist a "safety team" of employees from different positions in your establishment. Work with the team to assess the establishment's Hazard Communication Program. Focus on the following areas.

○ Hazardous chemical inventory. Is there a written inventory that indicates the chemicals by name and the places they are stored? Is the list current?

○ Labeling procedures. Are chemical containers labeled with the chemical's name, the manufacturers' name and address, and the potential hazards of the chemical?

○ MSDS. Are there MSDSs for each chemical stored at the establishment? Are these sheets accessible to all employees at all times while on the job?

○ Training. Are employees aware of the health hazards of all chemicals, and have they been trained to use them properly? Do they know what PPE is available, and can they use it properly? Do employees know where to find MSDSs, and do they know how to read them?

○ Written HAZCOM plan. Does the establishment have a written plan that describes how it will meet the requirements of the Hazard Communication Standard?

You may assign different "safety team" members to assess the current status in each of the above areas. Each team member should report his or her findings to the group so an action plan can be created to correct any deficiencies. The "safety team" should meet frequently (perhaps monthly) to assess how well the establishment is meeting the requirements of the HAZCOM program. "Safety team" members should receive an appropriate reward for their effort.

4. Pest Prevention Inspection

Purpose: *To identify areas in your facility where pests may enter, and to determine steps that should be taken to prevent entry.*

Directions: Conduct a thorough inspection of your entire facility. If possible, ask your PCO to perform the inspection with you. Pay particular attention to the following areas:

○ Doors, windows, and vents   ○ Floors and walls

○ Pipes exiting the building   ○ The building's foundation

○ Receiving areas (Note: Review your inspection practices to make sure you are inspecting for signs of infestation in incoming deliveries.)

Once possible entry points have been identified, a plan should be outlined for correcting these problems. While conducting the inspection, look for possible sources of food and shelter for potential pests. Check the way you currently clean the facility, store foods and supplies, and store and dispose of garbage and recyclables. Identify ways in which you might improve your current practices.

MULTIPLE-CHOICE STUDY QUESTIONS

1. You are opening a restaurant in an area not served by a city water supply. Which of the following is not a safe source of drinking water for your restaurant?
   A. A private underground well
   B. A creek with clear water that runs through the property
   C. A fixed above-ground water storage tank
   D. An approved mobile water storage tank

2. Which of the following is usually an approved use of nonpotable water?
   A. Sanitizing the restroom walls and floors
   B. Laundering the kitchen linen supply
   C. Operating the fire-safety sprinkler system
   D. Cleaning the warewashing machine

3. Why does a grease trap need to be cleaned regularly?
   A. A blocked grease trap can cause an overflow.
   B. Grease in a water supply line can give potable water a bad taste.
   C. Grease in a fire-safety sprinkler system can corrode pipes and cause leaks.
   D. Grease in floor-washing water can leave a film on the floor and cause an accident.

4. What is the primary reason for protecting the water supply from contamination?
   A. Microorganisms in contaminated water can make people ill.
   B. Sediment in contaminated water can cause service pipes to leak.
   C. Rust in contaminated water can permanently stain foodservice towels.
   D. The bad odor of contaminated water can make people ill.

5. Which of the following will not prevent backflow?
   A. An air gap between the sink drain and the floor drain
   B. The space between the faucet and the flood rim of a sink
   C. A vacuum breaker
   D. A cross-connection

6. You are installing an exhaust fan hood over your stove. What do you need to do to ensure its safe operation?

   A. The condensation collected must go directly into a floor drain.
   B. The hot air must be purified before being exhausted to the outside.
   C. The filters and ductwork must be cleaned regularly.
   D. The foods cooked under the hood must be covered at all times.

7. Where should kitchen trash containers be cleaned?

   A. In the "scrape" sink of a commercial three-compartment sink
   B. In a designated area away from food-preparation areas
   C. In a special area equipped with a bleach bath
   D. Any place in the kitchen where there is a floor drain

8. Why must handwashing stations be located close to food-preparation areas?

   A. To allow the handwashing sinks to also serve for washing vegetables if necessary
   B. To make it easy for workers to wash their hands frequently while preparing food
   C. To give workers easy access to a supply of paper towels for food preparation
   D. To allow the floor drain in the food-preparation area to also serve the handwashing station

9. Which hand-drying method is not recommended for use in a handwashing station?

   A. Hot-air dryer
   B. Single-use paper towels
   C. A common hand towel
   D. A continuous-cloth towel system

10. Which cleaning procedure must be performed daily in an establishment?

    A. Washing and sanitizing milk dispensers
    B. Washing ceilings and light fixtures
    C. Sanitizing storage shelves
    D. Washing dining area windows

11. Your restaurant purchases concentrated sanitizing chemicals. You want to make a spray solution for use in sanitizing food-contact surfaces in the establishment. What should you do to ensure that you have a proper sanitizing solution?

   A. Compare the color of the solution to another solution of known strength.
   B. Try out the solution on a food-contact surface.
   C. Test the solution with a sanitizer test kit.
   D. Use very hot water in making the solution.

12. Your supervisor has asked you to sanitize the food-preparation areas of the kitchen. Which of the following is a proper procedure?

   A. Spray with a strong sanitizing solution.
   B. Wash with a detergent, rinse, then wipe with a sanitizing solution.
   C. Wipe with a sanitizing solution, then rinse with clean water and wipe dry.
   D. Scrape and brush off soil, then wipe with sanitizing solution.

13. Which of the following is not a proper step in cleaning and sanitizing a standing food mixer?

   A. Remove the detachable parts and wash them in the warewashing machine.
   B. Dry the detachable parts with a clean cloth and reassemble the machine.
   C. Clean the food and dirt from around the base of the mixer.
   D. Wash and rinse nondetachable food-contact surfaces. Wipe them with a sanitizing solution.

14. Your warewashing machine is not working, and you must wash some tableware by hand. What is the first thing you should do to your three-compartment manual warewashing sink?

   A. Fill the first sink with hot water and detergent.
   B. Clean and sanitize the sinks and drainboards.
   C. Prepare the sanitizing solution in the third sink.
   D. Place clean, fresh towels on the drainboard next to the third sink.

15. What is the purpose of a Material Safety Data Sheet (MSDS)?

    A. To protect restaurant customers from potentially dangerous food substances

    B. To provide information regarding the proper use of potentially dangerous tools in the kitchen

    C. To keep foodservice workers informed of the hazards associated with the chemicals they work with

    D. To detail the specifications of proper construction materials for establishments

16. Which of the following is not a sign of a rodent infestation?

    A. Shiny black droppings
    B. Scraps of paper and cloth gathered in the corner of a drawer
    C. Small holes in a wall
    D. A strong oily odor

17. Your manager has asked you to check how pesticides used at the establishment are being stored. What should you check for?

    A. The container should be labeled with full manufacturer's information

    B. The brand of pesticide should be approved for commercial use by OSHA

    C. The pesticide used must be effective against cockroaches
    D. The containers should be recyclable

18. Which of the following is not a measure for controlling pests?

    A. Sealing the space around a pipe that exits the building
    B. Setting up a spring trap in the corner of the room
    C. Placing a chemical bait trap in the dishwashing area
    D. Setting out a multi-catch trap behind the icemaker machine

19. If pesticides are used in an establishment, containers should be

    A. discarded in the Dumpster when empty.
    B. rinsed in a food-prep sink when empty.
    C. rinsed in a warewashing sink when empty.
    D. discarded according to manufacturer's regulations.

Section 11
Sanitation Regulation

Learning Essentials

After completing this section, you should be able to:

○ Identify the principles and procedures needed to comply with food-safety regulations.

○ Identify state and local regulatory agencies and regulations that require food-safety compliance.

○ Identify the proper procedures for guiding an inspector through the establishment.

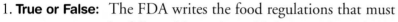

Knowledge

TEST YOUR FOOD-SAFETY KNOWLEDGE

1. **True or False:** The FDA writes the food regulations that must be followed by each establishment. *(See The Food Code, page 11-4.)*

2. **True or False:** Foodservice inspectors are generally employees of the Centers for Disease Control and Prevention (CDC). *(See Regulation, page 11-3.)*

3. **True or False:** You should ask to accompany the inspector during the inspection. *(See Foodservice Inspection Process, page 11-5.)*

4. **True or False:** Critical violations noted during an inspection should be corrected within one week of the inspection. *(See Foodservice Inspection Process, page 11-5.)*

5. **True or False:** The Model Food Code addresses the following topics on personnel: health, personal cleanliness, clothing, and practices. *(See The Food Code, page 11-4.)*

CONCEPTS

○ **USDA (United States Department of Agriculture):** The federal agency responsible for inspection and quality grading of meats, meat products, poultry, dairy products, egg and egg products, and fruits and vegetables shipped across state boundaries.

○ **FDA (Food and Drug Administration):** The federal agency that writes the Model Food Code based on input from the Conference of Food Protection (CFP). The FDA shares responsibility with the USDA for inspecting food processing plants to ensure standards of purity and wholesomeness and compliance with labeling requirements.

○ **Model Food Code:** Recommendations to assist state health departments in developing regulations for a foodservice inspection program.

○ **Health inspector:** A person who conducts foodservice inspections. Health Inspectors are also called sanitarians, health officials, and environmental health specialists. They are generally trained in food safety, sanitation, and in public health principles and methods.

INTRODUCTION

All establishments that serve the public must provide safe food and are subject to inspection. It does not matter whether there is a charge for the food, or whether the food is consumed on or off premises.

The purpose of an inspection program is as follows.

○ To evaluate the minimum sanitation and food-safety practices within the establishment

○ To protect the public's health by requiring establishments to provide food that is safe, uncontaminated, and properly presented

○ To convey new food-safety information to an establishment

○ To provide an establishment with a written report, noting deficiencies, so that the establishment can be brought into compliance with safe food practices

GOVERNMENT REGULATORY SYSTEM

In the United States, government control is exercised at three levels: federal, state, and local. At the federal level, the U.S. Department of Agriculture (USDA), the Food and Drug Administration (FDA), and the U.S. Public Health Service (USPHS) are directly involved in the inspection process.

USDA

○ This agency is responsible for inspection and quality grading of meats, meat products, poultry, dairy products, eggs and egg products, and fruits and vegetables shipped across state boundaries.

○ The USDA provides these services through the Food Safety and Inspection Service (FSIS) agency.

FDA

○ This agency writes recommendations for foodservice regulations (based on input from the Conference for Food Protection [CFP].) These recommendations are commonly known as the Model Food Code.

○ The FDA inspects foodservice operations that cross state borders (interstate establishments such as those on planes and trains, as well as food manufacturers and processors) because they overlap the jurisdictions of two or more states.

○ The FDA shares responsibility with the USDA for inspecting food-processing plants to ensure standards of purity and wholesomeness and compliance with labeling requirements. The USPHS inspects cruise ships that cross international borders.

Regulation

○ In the United States, most food regulations that affect restaurant and foodservice operations are written at the state level (except regulations for interstate or international establishments, which are determined at the federal level).

○ Each individual state decides whether to adopt the Model Food Code or some modified form of it.

○ State regulations may be enforced by local (city or county) or state health departments.

○ City, county, or state health inspectors (also called sanitarians, health officials, or environmental health specialists) conduct foodservice inspections in most states. They generally are trained in food safety, sanitation, and in public health principles and methods.

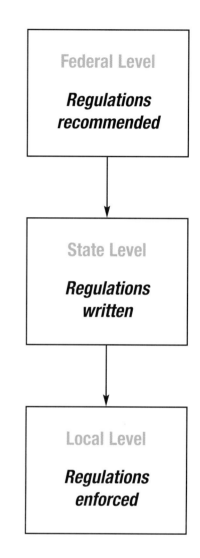

Federal Level

Regulations recommended

State Level

Regulations written

Local Level

Regulations enforced

THE FOOD CODE

The Model Food Code is written by the FDA and lists the government's recommendations for foodservice regulations. The Model Food Code is intended to assist state health departments in developing regulations for a foodservice inspection program. It is not an actual law. Although the FDA recommends adoption by the states, it cannot require it. Food codes are written very broadly, and generally cover the following areas.

○ Food handling and preparation: sources, receiving, storage, display, service, transportation

○ Personnel: health, personal cleanliness, clothing, practices

○ Equipment and utensils: materials, design, installation, storage

○ Cleaning and sanitizing of the facility and equipment

○ Utilities and services: water, sewage, plumbing, restrooms, waste disposal, integrated pest management (IPM)

○ Construction and maintenance of floors, walls, ceilings, lighting, ventilation, dressing rooms, locker areas, storage areas

○ Mobile and temporary foodservice units

○ Compliance procedures: foodservice inspections and enforcement actions

It is the responsibility of the manager to keep food safe and wholesome throughout the establishment at all times, regardless of the inspector or the inspection process.

FOODSERVICE INSPECTION PROCESS

Well-managed establishments will perform continuous self-inspections to protect food safety, in addition to the regular inspections performed by the health department. Establishments with high standards for sanitation and food safety consider health department inspections only a supplement to their self-inspection programs.

During health department inspections, the local health code serves as the inspector's guide. It is a good idea to keep a current copy of your local or state sanitation regulations. Regularly compare the code to procedures at your establishment, but remember that code requirements are only minimum standards to keep food safe.

The following suggestions will enable managers and operators to get the most out of food-safety inspections.

1. **Ask for identification.** Ask the purpose of the visit; make sure that you know whether it is a routine inspection, the result of a customer complaint, or for some other purpose.

2. **Cooperate.** Answer all of the inspector's questions to the best of your ability. Instruct employees to do the same. Explain to the inspector that you wish to accompany him or her during the inspection. This will encourage open communication and a good working relationship.

3. **Take notes.** As you accompany the inspector, make a note of any problem that is pointed out. Make it clear that you are willing to correct problems. If a deficiency can be corrected right away, do so, or tell the inspector when it can be corrected. If you believe that the inspector is incorrect about something, note what was commented upon. Then ask the inspector's supervisor for a second opinion.

4. **Keep the relationship professional.** Don't offer food or drink before, during, or after an inspection. This could be viewed as bribery.

5. **Prepare to provide records requested by the inspector.** If a request appears inappropriate, you can check with the inspector's supervisor or with your lawyer about limits on confidential information.

6. **Discuss violations and time frames for correction with the inspector.** Deficiencies and comments should be discussed in detail with the inspector. In order to make complete permanent corrections, you will need to know the exact nature of the violation, how it impacts food safety, how to correct it, and if the inspector will do any follow-up. Establishments are generally given a short amount of time (forty-eight hours or less) to correct major violations.

7. **Follow up.** Correct the problems. Determine why each problem occurred by evaluating sanitation procedures, the master cleaning schedule, and employee foodhandling practices and training. Establish new procedures or revise existing ones to correct the problem permanently.

Closure

A closure is issued when the health department feels that an establishment poses an immediate and substantial health hazard to the public. Examples of hazards that call for closure include:

○ A significant lack of refrigeration

○ A backup of sewage into the establishment itself or its water supply

○ An emergency, such as a building fire or flood

○ A serious infestation of insects or rodents

○ A long interruption of electrical or water service

SUMMARY

Today, government control regarding food safety in the United States is exercised at three levels: federal, state, and local. Some agencies at the federal level are directly involved in the inspection process. Most food regulations that affect restaurant and foodservice operations are written at the state level. Each state decides whether to adopt the Model Food Code or some modified form of it. State regulations may be enforced by local (city or county) or state health departments.

All establishments must follow standard food-safety practices critical to the safety and quality of the food served. An inspection system lets the establishment know how well it is following these practices. Well-managed establishments will perform continuous self-inspections to protect food safety, in addition to the regular inspections performed by the health department. Establishments with high standards for sanitation and food safety consider health department inspections only a supplement to their self-inspection programs.

TRAINING TIPS

Training Tips for the Classroom

1. Who Regulates What?

Objective: *After completing this activity, class participants should be able to identify the purpose of different regulatory and food-safety agencies.*

Directions: Create a handout to review the different regulatory and food-safety agencies.

Divide the sheet into two columns:

List of Agencies

- Food and Drug Administration
- United States Department of Agriculture
- Centers for Disease Control and Prevention
- Food Safety and Inspection Services
- National Marine Fisheries Services
- Environmental Protection Agency
- Local Health Department
- United States Public Health Service

Purpose of the Agency

Ask class participants to identify the purpose of each agency on the handout. Allow ten minutes for the class to complete the activity, and then discuss.

2. "If you're from the health department, where is your badge?"

Objective: *After completing this activity, class participants should be able to identify proper procedures for guiding a health inspector through an establishment.*

Directions: Conduct two role-plays.

Role-play 1: The wrong way

Select two class participants who have a flair for the dramatic, and ask one to play a health department inspector conducting a health inspection, and the other a restaurant manager. Ask the participants to refer to the proper procedures for guiding an inspector through an establishment as presented in Section 11, and do the opposite of what is suggested. Ask them to develop a dialogue that is adversarial, somewhat exaggerated, humorous, but realistic. Have them limit their role-play to five to eight minutes. After the role-play, discuss with the class what went wrong. Discuss the correct procedure for guiding an inspector through the establishment, as presented in Section 11.

Role-play 2: The right way

This role-play will demonstrate the proper way to guide an inspector through the establishment. Choose one class participant to act out this role-play with you. Depending on the class, you can choose to be the manager or the inspector in this activity. After the role-play discuss with the class what went "right."

Training Tips on the Job

1. Inspection Report Review and Preview

Purpose: *To evaluate the effectiveness of your food-safety system using the inspection requirements of your health department.*

Directions: Conduct an inspection of your establishment using the inspection form used by your health department. Don't be easy on yourself, don't make excuses, and don't make exceptions. Pretend that you are conducting the inspection of a competitor's establishment!

After the inspection, determine your food-safety score. You might solicit help from your health department in determining how to use and score the form. What corrective actions must be taken immediately, and what corrective actions can be taken later?

Before you actually make any corrections, you might ask your chef or assistant manager (or both) to conduct a similar inspection. How do the inspection reports compare? Do each of you see food safety from the same perspective? Chances are, your inspections will differ in some areas. Discuss these differences, and take corrective action.

MULTIPLE-CHOICE STUDY QUESTIONS

1. An establishment can be closed for all of the following reasons except
   A. a significant lack of refrigeration in the establishment.
   B. a backup of sewage in the establishment.
   C. a serious infestation of insects or rodents in the establishment.
   D. a minor violation in the establishment that was not corrected within twenty-four hours.

2. Which of the following is a goal of the food-safety inspection process?

   A. To evaluate the sanitation and food-safety practices within the establishment

   B. To protect the public's health

   C. To convey new food-safety information to establishments

   D. All of the above

3. A person shows up at a restaurant claiming to be a health inspector. What should the manager do?

   A. Ask to see identification.

   B. Ask to see an inspection warrant.

   C. Ask for a hearing to determine if the inspection is necessary.

   D. Ask for a twenty-four-hour postponement to prepare for the inspection.

4. Which of the following agencies enforce food safety in a restaurant?

   A. The FDA

   B. The Centers for Disease Control and Prevention (CDC)

   C. State or local health departments

   D. The USDA

5. Violations noted on the inspection report should be

   A. discussed in detail with the inspector.

   B. corrected within forty-eight hours or less if they are critical.

   C. explored to determine why they occurred.

   D. all of the above.

6. The responsibility for the sanitary operation of an establishment rests with

   A. the state health department.

   B. the manager/operator.

   C. the health inspector.

   D. the FDA.

Index